Introduction to
CHEMICAL PROCESS

Fundamentals & Design

Third Edition

Kenneth A. Solen
John N. Harb
Brigham Young University

The McGraw-Hill Companies, Inc.
Primis Custom Publishing

New York St. Louis San Francisco Auckland Bogotá
Caracas Lisbon London Madrid Mexico Milan Montreal
New Delhi Paris San Juan Singapore Sydney Tokyo Toronto

McGraw·Hill

A Division of The McGraw·Hill Companies

**Introduction to Chemical Process
Fundamentals & Design**

McGraw-Hill's Primis Custom Series consists of products that are produced from camera-ready copy. Peer review, class testing, and accuracy are primarily the responsibility of the author(s).

67890 QSR QSR 0987654321

ISBN 0-07-154054-7

Editor: Julie Kehrwald
Cover Designer: Mark Anderson
Printer/Binder: Quebecor Printing Dubuque, Inc.

To our wives, Linda and Ruth,
for their love and support.

Preface

To the Student

Chemical engineers are making important contributions to their companies and to society in an extremely broad range of products which span from microchips to potato chips. This book was written to introduce you to the exciting world of chemical engineering. To do this, you will be led step-by-step through a chemical engineering project similar to one that a professional chemical engineer might work on, and you will be introduced to many of the important aspects of chemical engineering along the way. At each step, you will learn something new about chemical engineering, as well as how that piece of the puzzle fits into the big picture.

Of importance is that this book is not just a qualitative discussion of the discipline. It actually teaches you how to perform some meaningful chemical engineering calculations and to make engineering decisions. A few of the skills you will acquire include how to construct material balances, calculate the work required from a pump, design a chemical reactor, and size a heat exchanger. You will also be introduced to engineering economics, computer tools, and principles of teamwork. By the end of the book you should have enough information to decide whether chemical engineering is a good field for you. You will also have an overview of the chemical engineering discipline, which will provide a foundation for future course work. Finally, it is our hope that you will have learned much and enjoyed the experience.

To the Instructor

Many freshman students considering whether to major in chemical engineering have very little information to assist them with that decision. Even at schools which offer a general freshman engineering class, the treatment of chemical engineering in such courses is too brief to provide an adequate picture of the discipline. Furthermore, students who major in chemical engineering often have very little idea of what the discipline is all about while taking their sophomore and junior courses, thus missing some of the learning that would take place if the student could recognize the connections between the material in those courses and its connection to the broader picture of the discipline (such a perspective, if obtained at all, often doesn't develop until the senior capstone course). Finally, even recognizing such needs, departments of chemical engineering may be reluctant to develop extensive introductory or freshman courses, because 1) heavy faculty commitments discourage the addition of more courses to be taught, 2) only a fraction of the students enrolled as freshman chemical engineering majors actually graduate in that field, and 3) the chemical engineering curriculum is already very full and bulging against the constraints of graduation time.

Emanating from the observations described above, this text is designed for an introductory course for first-year college students interested in chemical engineering. The goals of the book are 1) to provide a brief overview of the chemical engineering discipline at a level appropriate for beginning students and 2) to do so within a 2-credit, 1-semester course. The book leads the students from start to finish through a small design problem of a simple chemical process. In the first chapter, the student is asked to adopt the role of a chemical engineer on the job, and the design problem is introduced in the form of a memo from the student's supervisor. The next chapter provides brief descriptions of the chemical engineering discipline, chemical processes, and process flow diagrams. Throughout the remainder of the book, the students are led step-by-step through the design problem from initial problem definition to economic evaluation. At each step along the way, the concepts, principles, and equations necessary to accomplish the task at hand are introduced, an additional feature of the design is completed, and the process flow diagram is expanded to include that new feature. Thus, the student can see how the various aspects of chemical engineering integrate together to comprise an entire chemical process.

The goal of this book is to help the student experience chemical engineering to the fullest extent possible within the constraints of limited time and limited student background. In pursuit of that goal, it teaches the *freshmen* to solve quantitative problems, although *at a low level of complexity and within a scope that is narrow and well-defined.* These quantitative topics include material balances (reacting and non-reacting systems), fluid flow (including the sizing of pumps), mass transfer (diffusion and convection), chemical reactor design, heat transfer (including the design of heat exchangers), and engineering economics. As examples of the limited scope of these topics, the treatment of material balances for reacting systems is limited to single process units with one chemical reaction, and the treatment of fluid flow applications is restricted to the use of the mechanical energy balance where friction is mentioned, but friction factors and methods for determining friction losses are not introduced. Spreadsheets are also taught, and homework problems throughout the book give the students practice with this tool. In addition, a number of qualitative treatments are presented in the text, including chapters on problem solving, engineering teamwork, and process control. Finally, the students are given a few writing assignments to illustrate the important role of written communication in engineering.

Our intent was to design a book that would enhance learning for the beginning student. Consequently, we have tried to keep the text brief and to the point. Reading questions have been written for each chapter, with the intent that they be answered by the student after reading the assigned section and prior to discussion of the material in class. These questions include some which can be addressed directly and explicitly from material in the text and others that require additional thought and perception from the students. Nearly all of the key concepts in the text have been illustrated with examples. Also, we have included summary tables and charts which illustrate procedures and provide guidelines for specific types of problems. Homework problems are also included for each chapter.

Besides the design problem which forms the central story line throughout the book, there is also another design problem (comprising Chapter 14) for the students to do as a case study. This design problem requires the use of the skills taught throughout the semester to help students put all of the pieces together as they apply them to a new problem. We recommend that it be completed in teams and that it be introduced early enough (e.g. after Chapter 10) to allow time for completion by the end of the semester. In fact, Chapters 11-13 contain particularly light treatments of their subjects to allow students time to work on the case study.

You will find that students using this book begin to form an overview or conceptual "skeleton" of the chemical engineering discipline. As intended, this overview and experience provides a basis for the student to decide if chemical engineering matches his/her skills and interests. Those students who continue with the major are further benefited, because they recognize how their subsequent courses fit into the framework which is already in their minds. Their connection with their chemistry and physics courses is also enhanced as they recognize applications for those principles in their engineering field. While this introductory course in chemical engineering is not intended to replace, for example, the typical course in material and energy balances taught in the first or second year, our students have found that the introductory course leads to a richer experience in that later class, as well as in the junior and senior classes, by providing both perspective and background. With that perspective and background, those advanced students can focus on obtaining the technical depth or "meat" to fill in the "skeleton." Of particular value is that the students begin and end their educational program with a design orientation (most programs have a senior design course).

Ultimately, credit for this book goes to the students who inspired its conception and tested its effectiveness. The authors will be grateful for further feedback and suggestions as development of the book continues.

Table of Contents

Symbols

		Dimensions	Sample Units
c_A	molar concentration of Species A	moles/volume	*gmol/L*
C_p	heat capacity	energy/mass•temperature	*cal/g °C*
D_{AB}	binary diffusivity of Species A in Species B	area/time	*cm²/s*
E	energy	energy	*cal*
$\hat{E}$	specific energy	energy/mass	*cal/g*
E_a	activation energy of reaction	energy	*cal*
E_m	modulus of elasticity	force/area	*lb_f/in²*
g	gravitational acceleration	length/time²	*ft/s²*
h_m	mass-transfer coefficient	length/time	*ft/s*
h	heat-transfer coefficient	energy/time•area•temp	*cal/cm² s °C*
H	enthalpy	energy	*BTU*
$\hat{H}$	specific enthalpy	energy/mass	*BTU/lb_m*
k	thermal conductivity	energy/length•temperature	*cal/m °C*
k_r	reaction-rate constant	variable	
k_o	frequency factor	variable (same as for k_r)	
m	mass	mass	*kg*
m_A	mass of Species A	mass	*kg*
$\dot{m}$	mass flow rate	mass/time	*lb_m/hr*
$\dot{m}_A$	mass flow rate of Species A	mass/time	*lb_m/hr*
MW_A	molecular weight of Species A	mass/mole	*g/gmol*
n	moles	moles	*gmol*
n_A	moles of Species A	moles	*gmol*
$\dot{n}$	molar flow rate	moles/time	*lbmol/min*
$\dot{n}_A$	molar flow rate of Species A	moles/time	*lbmol/min*
$\dot{N}_A$	mass-convection rate of Species A	moles/time	*lbmol/min*
$\dot{N}_{A,x}$	mass-diffusion rate of Species A in x direction	moles/time	*lbmol/min*
P	pressure	force/area	*lb_f/in²*
$\dot{Q}$	rate of heat transfer across boundaries	energy/time	*J/s*
$\dot{Q}_{conv}$	rate of convection heat transfer	energy/time	*J/s*
$\dot{Q}_{cond,x}$	rate of conduction heat transfer in x direction	energy/time	*J/s*
$r_{cons,A}$	molar rate of consumption of Species A	moles/time	*kgmol/hr*
$r_{form,A}$	molar rate of formation of Species A	moles/time	*kgmol/hr*
$r_{react,A}$	rate of reaction	moles/time•volume	*gmol/s L*
R	universal gas constant	energy/mole•temperature	*cal/gmol °C*
$R_{cons,A}$	mass rate of consumption of Species A	mass/time	*kg/hr*
$R_{form,A}$	mass rate of formation of Species A	mass/time	*kg/hr*
T	temperature	temperature	*°C*
U	internal energy	energy	*cal*

$\hat{U}$	specific internal energy	energy/mass	cal/g
U_o	overall heat-transfer coefficient	energy/time•area•temp	$cal/cm^2 s\,°C$
v	velocity	length/time	m/s
V	volume	volume	cm^3
$\hat{V}$	specific volume	volume/mass	cm^3/g
$\dot{V}$	volumetric flow rate	volume/time	ft^3/hr
w_f	work of friction per mass of fluid	energy/mass	J/kg
w_s	shaft work on a fluid per mass of fluid	energy/mass	J/kg
W	weight	force	lb_f
$\dot{W}$	rate of work on a fluid	energy/time	J/hr
$\dot{W}_s$	rate of shaft work on a fluid	energy/time	J/hr
$\dot{W}_{PV}$	rate of flow work on a fluid	energy/time	J/hr
x	position along a linear coordinate	length	m
x_A	mass fraction of Species A	none	
y_A	mole fraction of Species A	none	
z	elevation	length	ft

Greek symbols

α	correction between $(v^2)_{avg}$ and v^2_{avg}	none	
ε	emissivity of a surface	none	
ε_x	strain	none	
ϕ	fractional tax rate	none	
v_A	stoichiometric coefficient of Species A	none	
ρ	density	mass/volume	g/cm^3
σ	Stephan-Boltzman constant	energy/time•area•temp^4	$W/m^2 K^4$
σ_x	stress	force/area	lb_f/in^2

Conversion Factors

| Acceleration | $1\ m/s^2 = 3.2808\ ft/s^2$ | $1\ ft/s^2 = 0.3048\ m/s^2$ |

Acceleration $1\ m/s^2 = 3.2808\ ft/s^2$ $1\ ft/s^2 = 0.3048\ m/s^2$

Area
$1\ cm^2 = 0.155\ in^2$ $1\ in^2 = 6.4516\ cm^2$
$1\ m^2 = 10.764\ ft^2$ $1\ ft^2 = 0.092903\ m^2$
  $1\ acre = 4046.9\ m^2$
  $1\ mi^2 = 2.59 \times 10^6\ m^2$

Density
$1\ g/cm^3 = 62.43\ lb_m/ft^3$ $1\ lb_m/ft^3 = 0.016019\ g/cm^3$
$1\ kg/m^3 = 0.06243\ lb_m/ft^3$ $1\ lb_m/ft^3 = 16.019\ kg/m^3$
  $1\ slug/ft^3 = 515.38\ kg/m^3$

Btu *kJ*
$\frac{}{1055\,kJ}$ $4.18\,\frac{}{cal}$

Energy
$1\ J = 0.7376\ ft\ lb_f$ $1\ ft\ lb_f = 1.3558\ J$
$1\ J = 9.478 \times 10^{-4}\ Btu$ $1\ Btu = 1055.0\ J = 778.1\ ft\ lb_f$
$1\ J = 2.778 \times 10^{-7}\ kW\ hr$ $1\ kW\ hr = 3.600 \times 10^6\ J$
$1\ J = 10^7\ ergs$ $1\ hp\ s = 550\ ft\ lb_f$
$1\ J = 0.2390\ cal$

Force
$1\ N = 0.22481\ lb_f$ $1\ lb_f = 4.4482\ N$
$1\ N = 10^5\ dynes$ $1\ ton = 2000\ lb_f$

Length
$1\ cm = 0.3937\ in$ $1\ in = 2.540\ cm$
$1\ m = 3.2808\ ft$ $1\ ft = 12\ in = 0.3048\ m$
$1\ km = 0.6214\ mi\ (statute)$ $1\ yd = 3\ ft$
$1\ km = 0.5400\ nmi\ (nautical)$ $1\ mi\ (statute) = 1.6093\ km = 5280\ ft$
  $1\ nmi\ (nautical) = 1.8520\ km$

Mass
$1\ g = 0.03527\ oz$ $1\ oz = 28.35\ g$
$1\ kg = 2.2046\ lb_m$ $1\ lb_m = 16\ oz = 0.45359\ kg$

Power
$1\ W = 0.7376\ ft\ lb_f/s$ $1\ ft\ lb_f/s = 1.3558\ W$
$1\ W = 9.478 \times 10^{-4}\ Btu/s$ $1\ Btu/s = 1055.0\ W = 778.1\ ft\ lb_f/s$
$1\ W = 1.341 \times 10^{-3}\ hp$ $1\ hp = 745.7\ W = 550\ ft\ lb_f/s$

Pressure
$1\ Pa = 1.450 \times 10^{-4}\ lb_f/in^2\ (psi)$ $1\ lb_f/in^2 = 6894.8\ Pa$
$1\ Torr = 1\ mm\ Hg\ (@\ 0°C)$ $1\ atm = 101,325\ Pa$
  $1\ atm = 760\ mm\ Hg\ (@\ 0°C)$
  $1\ atm = 14.696\ lb_f/in^2\ (psi)$
  $1\ atm = 33.9\ ft\ H_2O\ (@\ 4°C)$

Temperature $T(°C) = \frac{5}{9}\ [T(°F) - 32]$ $T(°F) = 1.8\ T(°C) + 32$

Viscosity $1\ cp = 6.7197 \times 10^{-4}\ lb_m/ft\ s$ $1\ lb_m/ft\ s = 1488.2\ cp$

Volume
$1\ cm^3 = 0.06102\ in^3$ $1\ in^3 = 16.387\ cm^3$
$1\ m^3 = 35.3145\ ft^3$ $1\ ft^3 = 0.028317\ m^3$
$1\ m^3 = 1000\ liters$ $1\ ft^3 = 7.4805\ gal$
$1\ m^3 = 264.17\ gal$ $1\ ft^3 = 28.317\ liters$
$1\ L = 0.26417\ gal$ $1\ gal = 3.785 \times 10^{-3}\ m^3 = 3.785\ L$

Volume Flow $1\ m^3/s = 15,850\ gal/min$ $1\ gal/min = 6.309 \times 10^{-5}\ m^3/s$
  $1\ gal/min = 2.228 \times 10^{-3}\ ft^3/s$
  $1\ ft^3/s = 448.8\ gal/min$

Defined Units

$$1 \, dyne \equiv 1 \, g \, cm/s^2$$

$$1 \, erg \equiv 1 \, dyne \, cm = 1 \, g \, cm^2/s^2$$

$$1 \, J \equiv 1 \, N \, m = 1 \, kg \, m^2/s^2$$

$$1 \, lb_f \equiv 32.174 \, lb_m \, ft/s^2$$

$$1 \, N \equiv 1 \, kg \, m/s^2$$

$$1 \, Pa \equiv 1 \, N/m^2 = 1 \, kg/m \, s^2$$

$$1 \, Poise \equiv 1 \, g/cm \, s$$

$$1 \, slug \equiv 1 \, lb_f \, s^2/ft$$

$$1 \, W \equiv 1 \, J/s$$

Atomic Weights of Selected Elements

ELEMENT	SYMBOL	ATOMIC WEIGHT	ELEMENT	SYMBOL	ATOMIC WEIGHT
Aluminum	Al	26.98	Manganese	Mn	54.94
Argon	A	39.99	Mercury	Hg	200.61
Barium	Ba	137.36	Molybdenum	Mo	95.95
Beryllium	Be	9.01	Neon	Ne	20.18
Boron	B	10.82	Nickel	Ni	58.71
Bromine	Br	79.92	Niobium	Nb	92.91
Cadmium	Cd	112.41	Nitrogen	N	14.01
Calcium	Ca	40.08	Oxygen	O	16.00
Carbon	C	12.01	Phosphorus	P	30.98
Cerium	Ce	140.13	Platinum	Pt	195.09
Cesium	Cs	132.91	Potassium	K	39.10
Chlorine	Cl	35.46	Scandium	Sc	44.96
Chromium	Cr	52.01	Silicon	Si	28.09
Cobalt	Co	58.94	Silver	Ag	107.88
Copper	Cu	63.54	Sodium	Na	22.99
Fluorine	F	19.00	Strontium	Sr	87.63
Germanium	Ge	72.60	Sulfur	S	32.07
Gold	Au	197.00	Tin	Sn	118.70
Helium	He	4.00	Titanium	Ti	47.90
Hydrogen	H	1.01	Tungsten	W	183.86
Iodine	I	126.91	Uranium	U	238.07
Iron	Fe	55.85	Vanadium	V	50.95
Lead	Pb	207.21	Zinc	Zn	65.38
Lithium	Li	6.94	Zirconium	Zr	91.22
Magnesium	Mg	24.32			

Some Values of the Universal Gas Constant, *R*

1.987 *cal/gmol K*

0.08206 *atm L/gmol K*

8.314 *J/gmol K*

CHAPTER 1

THE ASSIGNMENT

Welcome to this introductory text about chemical engineering. If you're like most beginning college students, you have very little idea of what the term "chemical engineering" means and what chemical engineers actually do. But there's good news — this book is designed to help answer those questions. Toward that end, this entire book is built around a chemical engineering problem and a role-playing scenario where you are a chemical engineer who is assigned by your company to solve an engineering problem.

Let's imagine that you are a chemical engineer for the ABC Chemical Company, which makes a valuable product but also makes hydrochloric acid (HCl) as a waste byproduct. Don't worry that you still don't know what chemical engineering is or what a chemical engineer does. You know enough to remember that an acid is a chemical which readily releases a hydrogen ion (H^+) and therefore reacts with many other substances. Continuing with our role-play, you have received the following memo from your supervisor:

ABC Chemical Company
Memorandum

From: Barbara Magelby, Supervisor, Chemical Process Group
To: <your name>

We've just received information indicating that the company which has been disposing of our HCl byproduct is not doing well. We anticipate that they will be going out of business in 6-12 months. This puts us in a very dangerous situation, since we can't operate very long without disposing of that waste. Our marketing people have tried to find a potential buyer for the acid, but the byproduct is apparently not at an appropriate concentration or purity to be valuable to anyone in our local area.

One possibility to consider is treatment of the waste in order to be able to dispose of it in the lake next to our company site. However, at this point, no engineering analysis has been conducted on this or any other strategy.

Your assignment is to propose a strategy and design (with a cost analysis) for safely and legally disposing of the acid waste.

Please keep me informed of your progress.

Even with very little background, you will recognize how important the problem described in this memo would be. Your company could not operate very long if it could not dispose of the acid waste product. The failure of your company would mean the loss of many jobs and would have a disastrous impact on many individuals and families. Society would also lose the products made by the company, which products help to make lives better. But even if it would keep your company operating, the acid cannot be simply "thrown away." The impact of hydrochloric acid on the environment is destructive. Governmental regulations prohibit its disposal unless it is rendered (and proven) non-destructive. Besides, your own conscience as an ethical and concerned citizen and professional would not allow you to participate in or condone any practice which is illegal, dangerous to other human beings, or harmful to the environment. Clearly, this is a problem which an engineer would take very seriously, and the solution to such a problem would bring great personal satisfaction.

In this book, this hypothetical assignment will be the basis for illustrating various aspects of the chemical engineering discipline. As we seek a solution to the problem, we will perform some relevant engineering calculations and even design some equipment. In addition, we will attempt to justify our proposal to the "company" through an economic analysis. By the time we reach the end of this book, you will have a better idea of the kinds of activities in which chemical engineers are engaged, the skills that are required for those activities, and how those activities and skills match up with your interests and abilities.

HOMEWORK PROBLEMS:

1. A major challenge in our society is balancing governmental protection of our environment with protection of industries that contribute to our economy. Using a computer word processor of your choice, write a short essay (at least one-half page, double spaced) describing the issues that you think should be considered in that balance.

2. Again, using a computer word processor of your choice, write a short essay (at least one-half page, double spaced) describing the ethical responsibilities that you think engineers have toward their employers and the responsibilities they have toward society. For example, what issues should an engineer consider if his/her company is violating environmental standards?

CHAPTER 2

WHAT IS CHEMICAL ENGINEERING?

In the presentation of our assignment in Chapter 1, we introduced the idea that you are working as a chemical engineer for a company which makes a valuable product by some chemical process and that we need to develop a strategy for dealing with an undesired byproduct. As we begin our introduction to chemical engineering, let's address the question posed in the title of this chapter:

What is Chemical Engineering?

Answer: The American Institute of Chemical Engineers (AIChE) has suggested the following definition:
Chemical Engineering is the profession in which a knowledge of mathematics, chemistry, and other natural sciences gained by study, experience, and practice is applied with judgment to develop economic ways of using materials and energy for the benefit of mankind.

Expanding upon this definition, chemical engineering uses all three of the basic physical sciences - chemistry, mathematics, and physics. This makes chemical engineering extremely **versatile**, since nearly all physical phenomena can be described by the combination of these three sciences. Because of this versatility, chemical engineers make valuable contributions in a very broad spectrum of fields, from food processing to semiconductor fabrication and from oil refining to artificial-organ development.

In general, chemical engineers convert chemicals from one form to another, either to produce modified materials for important applications or for generating power. In either case, this conversion operation takes place by means of a *chemical process*. It's time to develop this concept further by asking an important question:

What is a Chemical Process?

Answer: *A chemical process is a combination of steps in which starting materials are converted into desired products using equipment and conditions which facilitate that conversion.*

To explain this answer, let's go back to your experience with chemistry. You will remember that chemistry involves the use of chemical reactions to make a desired product. For example, we may be interested in making product "C" from chemicals "A" and "B" via the reaction:

$$A + B \rightarrow C \tag{2.1}$$

In your chemistry laboratory, you may have produced this reaction by pouring chemical "A" into a test tube and chemical "B" into another test tube (Figure 2.1). You may have then heated each test tube over a laboratory burner or heater to increase the temperatures of these two materials. The next step might have been to mix the two chemicals together so that they would react to form chemical "C." Finally, because other chemicals were present along with the "C" in the product mixture, you probably needed to separate "C" from the mixture by boiling it off from the mixture or by allowing it to settle to the bottom of a flask or by some other means.

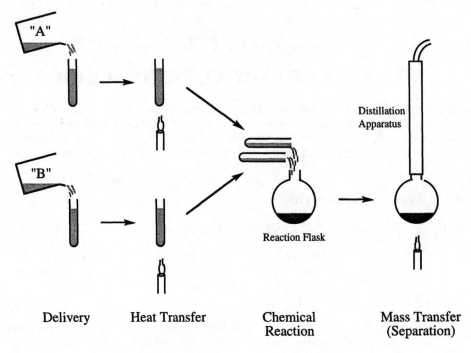

Figure 2.1 Schematic of a laboratory scheme for producing "C" from "A" and "B."

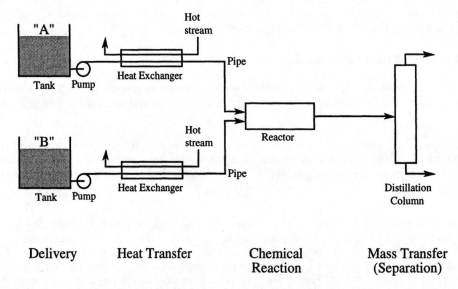

Figure 2.2 Schematic of an industrial **process** for producing "C" from "A" and "B."

In an industrial setting, we may wish to make large quantities (hundreds or maybe even thousands of pounds per hour) of chemical "C" continuously. Rather than hire hundreds of laboratory chemists with test tubes, we use a large-scale, automated chemical process. For example, the analog of the laboratory procedure described above might be to pump chemical "A" from a tank (Figure 2.2) through a pipeline containing a "heat-exchanger" (which transfers heat from a hotter stream to our colder stream of "A"). Similarly, we would probably pump "B" from another tank and through another heat exchanger. These "reactant" streams might then be brought together in a reactor with the needed temperature, pressure, and catalysts. It would be important that the reactor be designed so that the chemicals spend the needed amount of time inside the reactor before leaving, so that the reaction would proceed far enough (i.e. so that enough "A" and "B" would be converted to "C"). A final step might be to send the "product" stream to a continuous distillation (boiling) unit or to a settling tank or to another type of device which would separate "C" away from the other chemicals in the stream.

There are additional features of the industrial process not found in the laboratory scheme as well. These include equipment (usually including computers) to monitor important parameters (e.g. temperature, concentration, etc.) and to continuously adjust other variables to control the process. Materials and equipment for control of corrosion would also be present. Finally, the success or value of the process and the resulting product is ultimately affected by the economics of the process - i.e. the market value of the product versus the cost of producing that product.

A chemical process can be operated either as a *batch* process or a *continuous* process. These are defined as follows:

In a batch process, an allotment of starting material is introduced into the process, and a sequence of steps to treat that material is started and finished within a certain period of time, often within the same piece of equipment. The process is then interrupted, the processed material is removed, another allotment of the starting material is introduced, and the sequence of steps is repeated.

An example of a batch process would be loading materials into a reactor, followed by carrying out a reaction in the reactor, and then removing the final materials.

A continuous process operates without interruptions in the flows and reactions of the process. The starting material enters continuously, is usually subjected to various steps by moving from one piece of equipment to another, and exits the process continuously.

An example of a continuous process is a reactor where materials continually flow into and out of the reactor and the reaction proceeds as the material moves through the reactor.

In classifying processes, a second important distinction is whether a process is *"steady-state."* The following definition is useful:

A steady-state process is one in which none of the process characteristics (temperatures, flow rates, volumes, etc.) change with time.

A process which is not steady-state as defined above is termed *unsteady-state* or *transient*.

You will recognize that a batch process is clearly not a steady-state process. A continuous process may or may not be steady state, again depending upon whether any of the process characteristics vary with time. This should not be confused with the fact that the process variables may vary between different locations in the process. To clarify, when a steady-state process is observed at a certain point in time and then observed again a few minutes later, no change is seen. For example, in the process illustrated in Figure 2.2, the streams coming from both of the tanks pass through devices, called heat exchangers, which heat those streams so that the temperatures of the outlet streams are different from those of the inlet streams. If the process is steady-state, the inlet and outlet temperatures (and all other temperatures in the process) remain constant with time, in spite of the fact that those temperatures are not the same from one location to another. The same would be true of the pressures, chemical compositions, and other characteristics of the inlet and outlet (and other) streams in the process.

Example 2.1

A plant containing a reactor is being started up (being put into operation). In the *start-up phase*, with fluid flowing in all of the streams, the temperatures of the reactor and some of the streams are seen to be changing with time as they move toward the values at which they will eventually be held. The chemical composition of the material coming from the reactor is also changing with time in response to the changing reactor conditions. The start-up phase is over when the temperatures and compositions reach their desired values and no longer change with time. Classify the start-up phase and the period after the start-up phase in terms of being *batch* or *continuous* and *steady-state* or *unsteady-state*.

Solution: During both phases, the process is *continuous*, because the flows and operation continue without interruption or "starting over." During the start-up phase, the process is at *unsteady-state*, because some of the process parameters are changing with time. After the start-up phase, the process is at *steady-state*, because there are no changes with time.

Flowsheets

Processes are often represented by simplified flowsheets such as illustrated by Figure 2.2. Such flowsheets obviously leave out a great deal of detail, but they are useful because they show important sequences and relationships of steps in a chemical process and allow the engineer to easily visualize the process. They also provide important information about the process, such as the compositions, temperatures, and flow rates of process streams. Flowsheets are used by engineers and draftsmen involved in construction and maintenance (working on such things as piping, instrumentation, equipment design and plant layout), for the training of operating personnel, and for the preparation of operating manuals. The flowsheet represents the key documentation of the design and is the basis for comparison of actual operating performance with design specifications. Let's talk about three types of flowsheets:

         1) Block Diagrams

         2) Process Flow Diagrams

         3) Piping and Instrumentation Diagrams

Block Diagrams

Block diagrams provide a simple representation of a chemical process in which a box or block is used to represent either a single equipment item or a combination of equipment items which collectively accomplish one of the principal steps in the process. Such diagrams are especially useful at the early stages of process design before details have been determined. They can also be used to provide a simplified overview of the principal stages of a complex process. An example of a block diagram is given in Figure 2.3, which represents a nitric acid plant. Additional information such as stream flow rates and compositions may be shown on the diagram itself, or in a separate table.

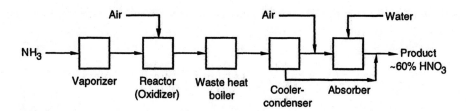

Figure 2.3 Block Diagram for a low-pressure process to produce nitric acid[1]

Process Flow Diagrams (PFD)

A Process Flow Diagram (PFD) provides more detail than a block diagram and is a standard method for documenting engineering designs. This type of diagram shows the arrangement and interconnection of all the major pieces of equipment and all flow streams, and the equipment is represented by symbols or icons which "look like" the actual equipment. Smaller equipment, such as pumps and valves may or may not be included in the PFD. Figure 2.4 shows the PFD for the same nitric acid plant represented in the block diagram in Figure 2.3. Each stream is identified with a number or letter, and the data for the streams are usually compiled in a stream table at the bottom of the flowsheet. The amount of information given in the stream table will vary but usually includes[1]:

1. Stream composition. Most commonly, this is given as:
   a) the flow rate of each individual chemical species or
   b) the percentage or fraction of each species in the stream
2. Total stream flow
3. Stream temperature
4. Normal operating pressure of the stream
5. The basis for the information in the table (e.g. in Figure 2.4, the indication in the top left corner of the table that the flow rates are in units of *kg/hr*)

Other information, such as physical property data, composition, etc., may also be added.

Standard symbols for various equipment items can be found in the American National Standards Institute (ANSI) publication on flowsheet design, but many companies have adopted their own symbols. Figure 2.5 shows some typical process equipment symbols.

Since a few homework problems in this book include the construction of some simple process flow diagrams, the following procedure is suggested for that construction:

1. Identify streams entering the process ("feed streams") and streams exiting the process ("product streams").
2. Identify key process steps and major equipment items needed for the process
3. Determine the symbol to be used for each major piece of equipment
4. Draw the symbols on the flow diagram and connect them with appropriate stream lines. The general flow of the diagram should be from left to right.
5. Label major pieces of equipment. Two common ways to do this are i) write the name of the equipment item near each symbol (as in Figure 2.3b) or ii) label the equipment with a one- or two-letter abbreviation indicating the type of equipment, followed by a number to indicate the individual equipment item (e.g., pumps could be labeled P-1, P-2, etc., and a heat exchanger could be labeled HX-101).
6. Label streams with a number and/or letter (e.g. 10, A, or 10A)
7. Include a stream table which contains information about each stream as indicated above.

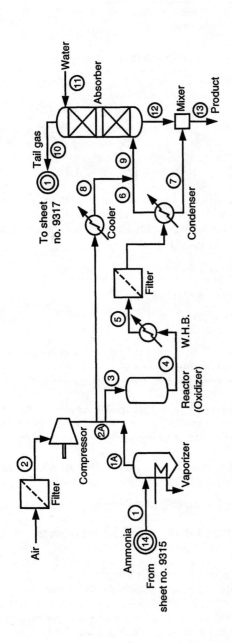

Flows kg/h	1	1A	2	2A	3	4	5	6	7	8	9	10	11	12	13	C & R Construction Inc.
Line no. Stream Component	Ammonia feed	Ammonia vapour	Filtered air	Oxidizer air	Oxidizer feed	Oxidizer outlet	W.H.B. outlet	Condenser gas	Condenser acid	Secondary air	Absorber feed	Tail (2) gas	Water feed	Absorber acid	Product acid	Nitric acid 60 percent
NH_3	731.0	731.0	—	—	731.0	Nil	—	—	—	—	—	—	—	—	—	100,000 t/y
O_2	—	—	3036.9	2628.2	2628.2	935.7	(935.7)[1]	275.2	Trace	408.7	683.9	371.5	—	Trace	Trace	Client BOP Chemicals
N_2	—	—	9990.8	8644.7	8644.7	8668.8	8668.8	8668.8	Trace	1346.1	10,014.7	10,014.7	—	Trace	Trace	SLIGO
NO	—	—	—	—	—	1238.4	(1238.4)[1]	202.5	—	—	202.5	21.9	—	Trace	Trace	Sheet no. 9316
NO_2	—	—	—	—	—	Trace	(?)[1]	967.2	—	—	967.2	(Trace)[1]	—	Trace	Trace	
HNO_3	—	—	—	—	—	Nil	Nil	—	850.6	—	—	—	—	1704.0	2544.6	
H_2O	—	—	—	Trace	—	1161.0	1161.0	29.4	1010.1	—	29.4	26.3	1376.9	1136.0	2146.0	
Total	731.0	731.0	13,027.7	11,272.9	12,003.9	12,003.9	12,003.9	10,143.1	1860.7	1754.8	11,897.7	10,434.4	1376.9	2840.0	4700.6	
Press bar	8	8	1	8	8	8	8	8	1	8	8	1	8	1	1	Dwg by Checked
Temp. °C	15	20	15	230	217	907	234	40	40	40	40	25	25	40	43	Date 25/7/1980

Figure 2.4 Process Flow Diagram (PFD) with stream table for a low-pressure nitric-acid process (adapted from ref. 1)

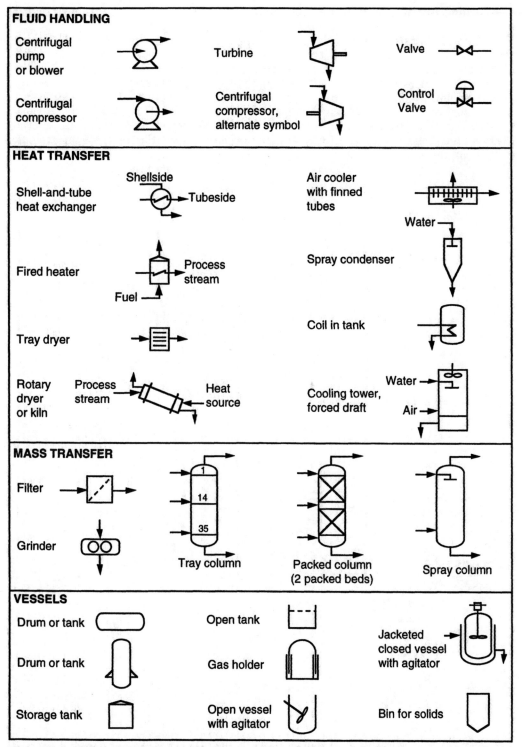

Figure 2.5 Typical symbols used in Process Flow and Piping and Instrumentation Diagrams

Piping and Instrumentation Diagrams (P&ID)

Piping and Instrumentation Diagrams (P&ID) represent the most detailed diagram of a process. P&IDs are prepared from the process flow diagram when sufficient detail is available and usually use the same numbers or letters as in the PFD to represent streams and equipment. However, the P&ID also includes the engineering details of equipment, instrumentation, piping, valves and fittings. Piping size, material specification, process instrumentation, and control lines are all shown on the P&ID. Utility (steam and high-pressure air) lines are also included on the diagram. All streams and utility lines entering and leaving the diagram are identified by source or destination. Figure 2.6 is an example of a small portion of such a diagram.

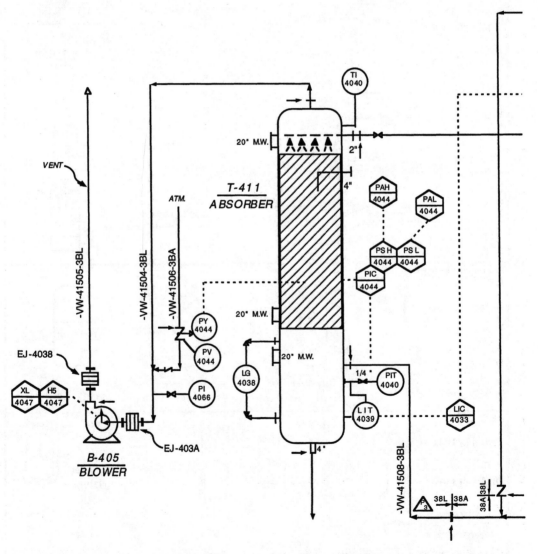

Figure 2.6 Example showing a portion of a Piping and Instrumentation Diagram (adapted from ref. 7)

Throughout this book, we will use both block diagrams and PFDs to represent chemical processes. However, P&IDs require technical calculations and details which are beyond the scope of this book and will not be used.

The Impact of Chemical Processing and Chemical Engineering

How much does chemical processing impact our lives? The answer may surprise you, because there is hardly a single moment of the day when you are not in contact with, surrounded by, or influenced by products which are a result of chemical processing. The paper you are looking at, the ink which forms the letters, and the pen you may be using to take notes are all a result of chemical processing. The clothes you wear, including the fabric and the dyes which produce the designs, are a result of chemical processing. The materials of the walls of the room around you and the paint on those walls were produced by chemical processing. Even the lunch you may have brought to school, including the paper bag in which it is carried, the sandwich bag and the bread of that sandwich, and the fertilizer and pesticide which helped to grow the fruit — all come from chemical processing. As you took your shower this morning, the soap, shampoo, deodorant, after-shave or perfume, and hair spray you may have used are all products of chemical processing. We haven't even mentioned the medication you may have taken to treat a medical problem or relieve a headache. We could go on and talk about the semiconductor components in the radio you listened to today or in the television you watched. The list is endless, because we live in an age when <u>virtually all materials which contact our lives are a result of chemical processing</u>.

To further understand the role of chemical processing in our modern society, we should recognize that we use many thousands of *finished products*, each of which is composed of many *intermediate products.* For example, the shampoo (a finished product) you used to wash your hair this morning probably contained the following kinds of intermediate products: the "soaps" (several) used for cleaning, nutrients (to nourish your hair, as promised in the advertising), pigments to give the shampoo a pleasing color, fragrances, surfactants, thickening agents, preservatives, etc. Each of these intermediate products was produced by chemical processing.

How did chemical processing come to play such an enormous role in our society? The answer really consists of thousands of stories of enterprising individuals and companies who identified a need. That need usually was associated with requirements for existing *finished products* but probably also included a vision about future products and needs. Thus, a company was formed or modified to produce the finished product. But the need for that product, in turn, generated a need for one or two or a dozen *intermediate products* which other companies were in a position to produce. And those intermediate products had to be produced from *raw feedstocks* — such as crude oil, biological byproducts, food stuffs, or ores — which had to be converted into more usable byproducts, and still other companies responded to that need. That story has been repeated many hundreds of times over the years, as the refinement of raw feedstocks, the production of intermediate products, and the design of finished products have developed together. One indication of that development is that the number of intermediate products or chemicals available on the industrial market has grown to a list of thousands. Table 2.1 lists some of the categories of intermediate products, where each category represents tens to hundreds of specific chemicals. The average consumer may not ever see most of the products listed in Table 2.1 as isolated products but uses them very frequently without being aware of them as components of finished products.

Another way of illustrating the vast influence of chemical engineering on our modern society is to summarize some of the historical achievements of this field. Table 2.2 briefly describes 10 such achievements, but you should understand that this is by no means an exhaustive list.

Table 2.1 Some Categories of Intermediate Products

Absorbents	Adhesives	Antiadhesives
Antioxidants	Catalysts	Cleaners
Coatings	Corrosion inhibitors	Combustion modifiers
Desiccants	Disinfectants	Drying agents
Dyes	Emulsifiers	Explosives
Fertilizers	Fillers	Flame retardants
Flavorings	Foods	Fragrances
Fuels	Herbicides	Industrial reagents
Insecticides	Lubricants	Nutritional supplements
Oxidizers	Paper products	Pesticides
Pharmacological agents	Pigments	Plasticizers
Polymeric feedstocks	Preservatives	Sealants
Solvents	Structural materials	Supports
Surfactants	Textiles	Thickeners
Thinners	UV Screeners	Wetting agents

Fundamental Topics

As we pointed out earlier in the chapter, chemical engineers use the principles of chemistry, physics, and mathematics to understand and describe the various phenomena that occur in a chemical process. A few of those phenomena were alluded to in the processes represented by Figures 2.2, 2.3, and 2.4. The fundamental physical phenomena of interest to chemical engineers can be grouped into the following topics:

1. **Fluid mechanics** (how fluids flow) so that the movement of fluids can be described and produced (e.g. in pump-pipeline systems)

2. **Heat transfer** (how heat transfers) so that the heating or cooling of chemical materials can be described and produced (e.g. in heat exchangers)

3. **Mass transfer** (how molecules move relative to each other) so that the mixing or separation of chemical species can be described and produced, using such strategies as

   distillation: boiling a mixture to preferentially remove some substances from the mixture which boil off in higher proportion than do other substances in that mixture

   evaporation: removing a substance from a mixture by allowing it to vaporize

   drying: removing (by evaporation) all liquid from a material

   filtration: separating solid particles from a suspending liquid or gas using a filter

   liquid-liquid transfer: (also called "extraction") transferring a substance from one liquid mixture to another liquid by contacting the two liquids (the two liquids must be immiscible, i.e. must not dissolve into one another)

Table 2.2 Some Great Achievements of Chemical Engineering (adapted from ref. 6)

Crude Oil Processing: Chemical engineers discovered ways to use crude oil by breaking down ("cracking") long carbon molecules into smaller ethylenes, propylenes, etc. These advancements opened the way for the development of many of the petrochemical products used today, including fuels, plastics, pharmaceuticals, etc.

Plastics: During the awakening of the "plastics" age in the 1930's, chemical engineers helped to develop Nylon, polyvinylchloride, acrylics, polystyrene, polyethylene, and polypropylene, followed more recently by aerospace plastics, composites, and laminates.

Synthetic Fibers: Beginning with the first extrusion of rayon fibers in 1910 and "nylon" in 1939, chemical engineers have helped to develop and manufacture many synthetic fibers, from blankets, to clothing, to upholstery, to carpets, to very strong (even "bullet-proof") fabric. Today, eight billion pounds of synthetic fibers are produced each year.

Synthetic Rubber: Chemical engineers helped to develop synthetic rubber (styrene-butadiene rubber or SBR) during World War II, when natural rubber was hard to obtain. Today, two-thirds of the 22 billion pounds of rubber produced annually is synthetic, and SBR still accounts for half of all synthetic rubber produced. Synthetic rubber is used in tires, gaskets and hoses, and in many consumer products, such as running shoes.

Gases from Air: Chemical engineers pioneered ways to process and use purified oxygen and nitrogen. Air is separated into these gases through "cryogenics" (liquefying the gases at very low temperatures). Nitrogen is used as an inert gas in many applications, while oxygen is used in many industrial processes and in hospitals. Chemical engineers also developed vacuum-insulated tank trucks for delivering liquefied gases.

Environmental Protection: Chemical engineers are seriously involved in environmental technology to clean up existing problems and to prevent future pollution. Chemical engineers have helped to develop catalytic converters for cars, double-hosed gasoline pumps, and modern jet engines. Scrubbers on smokestacks also help maintain air quality.

Food: Chemical engineers have helped to produce fertilizers, like phosphates and urea, and pesticides that protect crops. They are also at the forefront of efforts to improve food processing technology, such as freeze-drying and microwave processing.

Separation and Use of Isotopes: Chemical engineers developed the processes which separate isotopes from one another. Chemical engineers also played a major role in developing and using isotopes from fission in nuclear medicine as advanced diagnostic and treatment techniques. Other uses of radioisotopes include medical imaging and monitoring, biochemical research, archaeological dating, and production of nuclear energy.

Medicine: Because the body is essentially a chemical plant, chemical engineers have worked hand-in-hand with physicians to develop medicines, replacement parts, and physiologic support products. The resulting biomedical technology has led to improved clinical care, high-tech diagnostic and therapeutic devices, new wonder drugs, and artificial organs.

Antibiotics: Chemical engineers have developed ways to mass produce antibiotic drugs, For example, with penicillin (discovered in 1929), research led to a thousand-fold increase in yield through mutation and through development of methods for "brewing" penicillin in huge tanks. These and similar developments have resulted in low-cost, widely-available "wonder drugs" including penicillin, streptomycin, erythromycin, and many more.

gas-liquid transfer: (also called "absorption") transferring a substance from a gas mixture to a liquid by contacting the gas with a liquid into which the desired substance preferentially dissolves

4. **Reaction engineering** (including how fast chemical reactions occur) so that the quantity of material converted by reactions can be predicted or achieved by design (e.g. in reactors)

5. **Process control** (how to adjust process variables to hold key parameters within specification) to maximize product quality and lower operating costs

6. **Materials and corrosion** (how materials respond to mechanical stress and/or chemical attack) so that materials can be selected for products and/or to minimize the degradation of process equipment

7. **Economics** (costs of various design and operating options) so that choices may be made between options

The understanding of these topics is used to develop strategies and/or to design equipment which combine these individual steps into a complete process. For example, some chemical engineers design entire chemical plants containing thousands of feet of pipe and many pumps, valves, tanks, columns, heat exchangers, reactors, and computer systems. Other chemical engineers predict the creation, movement, and loss of environmental pollutants, along with the effects of various proposed measures to reduce such pollutants. Still other chemical engineers search for more effective and cheaper ways to make vital pharmaceutical agents for medical purposes.

Professional Activities

The activities of the chemical engineering profession fit into several general areas, which are briefly summarized as follows:

Process Development Research: These chemical engineers develop and refine methods to produce chemical products with an existing market or to produce new products. They do this with mathematical simulations, laboratory experiments, and pilot-plant studies.

Technical Chemical Sales: Technical sales representatives from a company work with existing or potential purchasers of the company's product(s) to assess the needs of those customers and to match the products of their companies with those needs.

Process Engineering: Process engineers deal with the operations surrounding the production of chemicals in existing facilities, including quality control, operations management, trouble shooting, safety, and maintenance.

Plant Design and Construction: Design engineers formulate the specifications of equipment and plants to implement a desired chemical process and direct the construction and start up of those facilities.

Environmental Engineering: Environmental engineers monitor the compliance of process operations to environmental regulations and implement measures and processes to prevent violation of those standards and/or to clean up hazardous waste.

Finally, it's important to note that <u>chemical engineers work with people</u> and must have effective communication skills. Their technical work is usually performed in teams with their professional colleagues. In such teams, chemical engineers must be able to share technical information, convey relevant reasoning, and influence important decisions. Many also interact with non-technical personnel, ranging from business and government representatives to the general public, and it is vital that such engineers also communicate effectively and provide leadership in addressing the concerns of various groups within our society.

Summary

This chapter has presented brief definitions of the chemical engineering profession and of a chemical process. Some of the topics and professional activities of chemical engineering were also briefly described. This introductory textbook is designed to guide your study of these topics in order to increase your understanding of their concepts and your perception of the nature of this profession.

References

1. Sinnott, R.K., *Chemical Engineering Vol. 6: An Introduction to Chemical Engineering Design*, Oxford: Pergamon Press, 1989.

2. Couper, J.R., *Class Notes*, Personal Communication, March 1995.

3. *Chemical Engineering Orientation*, 2nd ed., Sugar Land, TX: Career Focus, Inc., 1993.

4. *From Microchips to Potato Chips: Chemical Engineers Make a Difference*, New York: American Institute of Chemical Engineers, 1991.

5. Chenier, P.J., *Survey of Industrial Chemistry*, 2nd Revised Edition, New York: VCH Publishers, Inc., 1992.

6. *Ten Greatest Achievements of Chemical Engineering*, New York: American Institute of Chemical Engineers, 1992.

7. Baasel, W.D., *Preliminary Chemical Engineering Plant Design*, 2nd ed., New York: Van Nostrand Reinhold, 1990.

8. Shreve, N.R. and J.A. Brink, Jr., *Chemical Process Industries*, New York, McGraw-Hill Book Co., 1977.

READING QUESTIONS:

1. Which three physical sciences are used in chemical engineering?

2. In the definition of a chemical process, what is the purpose of the equipment and conditions used in the process?

3. In our example of making chemical "C" from "A" and "B," we formulated an industrial process to replace a laboratory scheme. What equipment in the industrial process replaced each of the following laboratory equipment items?

> test tubes
> laboratory burner
> reaction vessel
> distillation apparatus

4. How does a continuous process differ from a batch process?

5. What distinguishes a process as being steady state?

6. Name the seven kinds of fundamental topics which chemical engineers must learn about.

7. How does a PFD differ from a block diagram?

8. Beginning with an idea for producing a chemical product, briefly describe various stages of development which might be used to make the product and provide it to the end user. Indicate how the various activities of chemical engineering (Process Development Research, Technical Chemical Sales, Process Engineering, Plant Design and Construction, and Environmental Engineering) might be involved in those stages.

HOMEWORK PROBLEMS:

1. From your home or apartment, select one of each of the following kinds of items and, from the information on the label or container, write down all of the "ingredients" (intermediate products): candy bar, deodorant, laundry detergent, pain medication.

2. List common examples from everyday life for all of the following fundamental operations: fluid mechanics, heat transfer, evaporation or drying, and filtration.

3. Classify the following as either <u>batch</u> or <u>continuous</u> processes and also indicate whether each is a <u>steady-state</u> or an <u>unsteady-state</u> process.

    a. A "surge tank" is used when a liquid is coming from one part of a process at a variable rate, and we want to provide a reservoir of that liquid to feed another part of the process. Thus, a surge tank continuously receives liquid from an incoming stream and also loses that same liquid continuously in an outgoing stream. Because the flow rates of the incoming and outgoing streams are changing with time, the volume in the tank also changes with time.

    b. We bake a cake by mixing together the ingredients in a cake pan, placing the pan and mixture in an oven for a prescribed amount of time, and then removing the cake to cool down.

    c. A company produces latex paint base by mixing together the ingredients for the paint. All flow rates are held constant to maintain the proper ratio of ingredients. Working around the clock, the company makes approximately 800 *gallons* of paint every 24 *hours*.

4. Hydrogen gas is a valuable product, because it is used as a feedstock (starting material) for many chemical processes. A common way to produce high-purity hydrogen gas is by reaction of propane gas with steam using the following scheme:

    A. The propane gas is first sent to a Desulfurizer to remove any sulfur present in the propane gas, because the sulfur would poison catalysts in later process steps.

B. Steam is added to the desulfurized propane, and the combined gas is then sent to a Reforming Furnace (a fired heater) (1500°F) to produce the "reforming" reaction:

$$C_3H_8 + 3H_2O \rightarrow 3CO + 7H_2$$

C. More steam is added to the gas mixture leaving the Reforming Furnace, and the combined gas goes to a CO Converter, where the carbon monoxide in the mixture is converted as follows:

$$CO + H_2O \rightarrow CO_2 + H_2$$

D. The gas mixture from the CO Converter enters the CO_2 Absorber, where most of the CO_2 in the mixture is absorbed into an amine solution.

E. The gas mixture from the CO_2 Absorber now contains H_2 with traces of CO and CO_2. The last traces of CO and CO_2 are converted to methane in a Methanator as follows:

$$CO + 3H_2 \rightarrow CH_4 + H_2O$$
$$CO_2 + 4H_2 \rightarrow CH_4 + 2H_2O$$

a.   Construct a Block Diagram for the process described above.

b.   Construct a pictorial Process Flow Diagram (without the stream table) using the symbols given in Figure 2.5. The following additional information will be helpful:

•Liquid propane will be fed from a Propane Tank

•The propane leaving the Propane Tank is vaporized via a shell-and-tube heat exchanger in which steam is used on the tube side (see the next bullet)

•A shell-and-tube heat exchanger (as you will learn later) is a large cylinder (shell) through which a number of tubes pass. One fluid flows inside the tubes ("tube side"), and the other fluid flows outside the tubes but inside the outer cylinder ("shell side"), and the streams don't mix. In the symbol for a shell-and-tube heat exchanger, a line representing the tube side passes through the circle (which represents the shell or outer cylinder). Other lines stop at the boundary of the circle to represent the shell-side fluid entering and leaving the shell. The orientations and directions of the lines and arrows are not critical.

•The Reforming Furnace is a fired heater

•The Desulfurizer, CO Converter, CO_2 Absorber, and Methanator are packed columns with one bed of packing each, and the process gas enters the bottom and exits the top. In the case of the CO_2 Absorber, amine solution enters the top ("amine solution in") and exits the bottom ("amine solution out"), where the source and destination of the amine solution streams will not be indicated.

•For the CO_2 Absorber, the Absorber inlet stream is cooled by 2 shell-and-tube heat exchangers in series. In the first exchanger, the Absorber inlet stream flows through the shell side, while cool outlet gas is looped back from the Absorber to flow through the tube side before continuing on its way to the Methanator. In the second exchanger, the Absorber inlet stream again flows through the shell side, while water is used as the tube-side coolant.

•The gas leaving the Methanator is cooled again by water in a shell-and-tube exchanger (the gas on the shell side), and the diagram should label the cooled stream as "Purified hydrogen to storage."

5. A common and important process is the manufacture of gelatin for food, pharmaceuticals, photographic film, and various technical applications. The chemistry is the simple hydration of collagen from animal bones or skins:

$$C_{102}H_{149}N_{31}O_{38} + H_2O \rightarrow C_{102}H_{151}N_{31}O_{39}$$

collagen   water   gelatin

Bones must be pretreated with steam to remove the grease, then crushed into small particles, and then sent to a series of acid-wash steps to remove calcium phosphate and other mineral matter. The remaining collagen then goes to long storage (1 month or more) in lime to remove soluble proteins before finally going to the reactor and purification processes. An abbreviated process flow diagram (without a stream table) appears below.

For your interest, to produce 1 *ton* of gelatin requires approximately 3 *tons* of bones, 1 *ton* of hydrochloric acid, 3/4 *tons* of lime, and 400 *lbs.* of steam.

For each of the following operations in this process, which of the first four chemical engineering phenomena (fluid mechanics, heat transfer, mass transfer, and reaction engineering) do you think are important parts of the operation?

a) **Cooker:** Steam is used to heat the bones and cause the grease to flow more easily. The stream of grease is caused to flow away from the bones.

b) **Acid Wash:** A stream of acid is brought in and mixed with the bone particles. The acid reacts with the solid material of the bones to break down that material and release the calcium phosphate and other minerals. The acid stream also carries away the minerals as it leaves the process.

c) **Dryer:** Steam is brought into the compartment where strips of gelatin are laying on trays. The steam heats the compartment until the gelatin is completely dry.

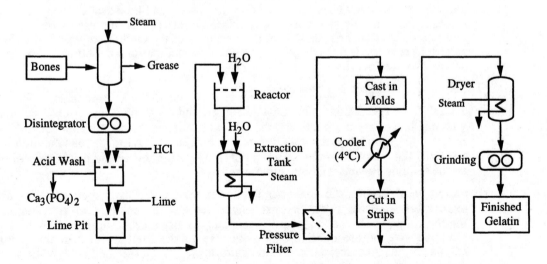

Process for Gelatin Manufacture (adapted from reference 8)

CHAPTER 3

SOLVING ENGINEERING PROBLEMS
(WHAT SHALL WE DO?)

Strategies for Solving Problems

As mentioned in Chapter 2, chemical engineers use chemical processes to produce a variety of useful products. The successful design and operation of those processes requires that chemical engineers solve many different kinds of problems. Typically, the kinds of problems that must be solved are initially poorly defined, meaning that a problem is evident (e.g. the process is not working), but the most decisive questions or issues are not yet apparent. Problems encountered by chemical engineers also have several possible solutions. For example, the problem introduced in Chapter 1 (the company which was disposing of our HCl is going out of business) might be solved in a number of ways. Once problems are more clearly defined and one or more solution strategies are selected for analysis, the engineer then performs specific calculations to address well-defined questions. Many of the homework problems in this book are representative of such engineering calculations.

Now, it's time to introduce some steps for solving any engineering problem and to apply them to the problem which was described in Chapter 1. Such steps can be defined and grouped in a variety of ways, but the same elements are found in any such construction. One list of steps would be as follows:
1. Define the problem
2. List possible solutions
3. Evaluate and rank the possible solutions
4. Develop a detailed plan for the most attractive solution(s)
5. Re-evaluate the plan to check desirability
6. Implement the plan
7. Check the results

Now, let's describe each of these steps and then apply them to the problem at hand.

1. Define the problem

Very often a problem goes unsolved because it has not been defined clearly and correctly. This is illustrated by the following example:

A chemical engineer working for an oil company was assigned to solve a problem associated with a piece of equipment which the company had recently purchased and installed but which did not perform according to expectations. The equipment was a heat exchanger, a device for transferring heat from a hot stream to a cold stream, and was purchased to heat a particular stream of liquid to a certain temperature. When the young engineer investigated the situation he learned the following:

The oil company had ordered the heat exchanger from a reputable specialized company which only made such devices. As part of the order, the amount of heat that the exchanger had to be able to transfer had been specified by an engineer at the oil company. Once constructed, delivered, and installed, the heat exchanger did not raise the temperature of the targeted liquid stream to anywhere near the desired value. Engineers at the oil company had been debating among themselves for several months about the reasons for the inadequate performance, with some arguing that the heat exchanger had been constructed improperly and others insisting that it had been installed incorrectly, with both sides proposing elaborate theories to support their arguments.

When the young engineer collected temperature and flow-rate information for the streams going through the exchanger, his calculations indicated that the exchanger was transferring the amount of heat specified in the original order. Indeed, the amount of heat <u>needed</u> to raise the liquid stream to the desired temperature was larger than had been specified in the order. For months, the engineers had been defining the problem as a malfunction of the heat exchanger. The more correct definition, that the temperature of the liquid stream was not being raised sufficiently, would have allowed them to consider the possibility that the heat exchanger had been incorrectly specified in the original order.

For our problem, we are looking for the optimum strategy for dealing with the acid waste from our process. From our company records, we learn that the average acid stream flow rate is 11,600 *L/hr* and that the average HCl concentration in the waste stream is 0.014 *M*.

2. List possible solutions

In generating possible solutions, we need to keep an open mind and not discard ideas too quickly. Sometimes, strategies which seem impractical at first turn out to be better than originally thought or become springboards to better ideas. Therefore, a list of many possible ideas should be written down before evaluating any of them.

For our problem, how many possible solutions can you think of? Let's list a few:
a. Change our company process so that the acid is not produced
b. Contract with another independent company to take the acid away
c. Build giant holding tanks to store the acid for 10 years
d. Discharge the acid to an evaporation pond built on the company site
e. Treat and discharge the acid into the lake next to the company site

3. Evaluate and rank the possible solutions

Full evaluation of some of these strategies would require consideration of many factors ranging from construction costs to governmental fees and could actually involve days, weeks, or months to complete. For the sake of this exercise, only brief arguments and conclusions will be presented to illustrate the evaluation process.

a. <u>Change the process so acid is not produced</u>: This may be possible, but alternate processes are usually not known or are extremely expensive. For our present operation, let's say that no alternate process is known.

b. <u>Contract with another company to take the acid away</u>: An independent company would have to treat and dispose of the acid just as our company would have to do. In addition, transportation costs and profit for that company would be added onto the charge we would have to pay. Thus, it would seem that we can do it more cheaply. This would not be true, however, if the independent company already had equipment for doing the necessary treatment. For this analysis, let's say that no such company can be identified.

c. Build holding tanks for 10 years of storage: The required volume of such tanks would be

$$11,600 \ L/hr \times 24 \ hrs/day \times 365 \ days/yr \times 10 \ yrs = 1.0 \text{ billion } L$$

(notice how that calculation was made). To estimate how many tanks would be needed, suppose that each tank was cylindrical and was 10 m in diameter and 5 m high. The volume occupied by each tank would be

$$\text{Volume} = \frac{\pi}{4} D^2 H = \frac{\pi}{4}(10m)^2(5m) = 393m^3$$

Knowing that there are 1000 L in a cubic meter, you should be able to calculate that 2,587 tanks would be needed. We can prepare a cost estimate for building and maintaining this many storage tanks, but our intuition tells us that this would be prohibitively expensive. In addition, we would still have to do something after 10 years to allow us to keep operating, such as building still more tanks.

d. Discharge the acid to an evaporation pond: Evaporation ponds are useful for concentrating waste solutions. We would need to construct the ponds, making sure that no acid leaked through into the ground water. Land will have to be available, since these ponds will be large (remember we have to evaporate 11,600 L of water per hour). If that land is distant from the company site, transportation costs will need to be included in the cost estimates. If the temperature is near or below freezing for a significant fraction of the year, the evaporation rate would be low and unpredictable during that time, and the ponds would need to accumulate at least 6 months of waste (50 million L) and to evaporate at twice the yearly average rate (23,200 L/hr) during the warmer months. A study of evaporation rates and processes suggests that the construction of ponds large enough for this process would be excessively expensive. In addition, the state environmental agencies would require a study to ensure that the amount of acid carried into the air with the evaporating water would not exceed maximum limits. Finally, we would eventually need to dispose of the concentrated acid.

e. Treat and discharge the acid to the lake: If the acid was added to the lake without treatment, the change in acidity of the lake would kill both the plant life and the fish in the lake. Such a result would be contrary to your own feelings of environmental responsibility and would violate the regulations of the U.S. Environmental Protection Agency (EPA) and the state water quality standards, resulting in heavy fines and legal prosecution. Therefore, any plan to discharge the stream into the lake must include treatment of the stream, such as neutralizing the stream with a basic solution.

An additional consideration would be the temperature of the final discharge to the lake, since the waste stream is usually much warmer than the lake. Plant and fish life would both be affected by the continual addition of excessively-warm water. The average temperature of the entire lake may rise. More importantly, the local temperature in that part of the lake where the discharge takes place would rise significantly. Thus, the process will need to include cooling the stream to a target temperature.

This process of neutralization and cooling might be similar to the strategy used by the company which has been disposing of the acid up to now. A process design would need to be established, and the cost of construction and operation would need to be estimated. One advantage of this plan is that the operating expense of this on-site process should be less than the fees paid to the previous disposal company, and the savings may offset the cost of building the new facility.

4. Develop a detailed plan for the most attractive solution(s)

The analysis of the various options illustrated above is brief, and a more in-depth and exhaustive analysis would normally be conducted. Let's suppose that a preliminary analysis suggests that option "e" is the best option. Continuing the scenario in which you imagine yourself as the engineer assigned to make this decision, let's suppose that you have selected the strategy of neutralizing the acid by addition of sodium hydroxide to the waste stream. Sodium hydroxide is an example of a class of chemicals that we call "bases," which readily release a hydroxide group (OH^-), and bases and acids react together to form water. In this case, the reaction between the sodium hydroxide and the hydrochloric acid is

$$HCl + NaOH \rightarrow H_2O + NaCl \tag{3.1}$$

You will recognize that the sodium chloride (NaCl) is common table salt. Adding it to the lake will increase the salinity of the lake. Knowing that salt falls under the category of "dissolved solids," we do a little homework to find the maximum dissolved solids allowed by state regulations. For example, in Utah, those regulations are found in the document "Standards of Quality for Waters of the State," which is part of section R317-2 of the Utah Administrative Code. In that document, we learn that the upper limit for dissolved solids in water which may have agricultural applications is 1200 *mg/L*, which is the same as 1.2 *g/L*. From the atomic weights given in the front of the book, the molecular weight of sodium chloride is 23.0 + 35.5 = 58.5 *g/gmol*. Therefore, the final concentration cannot exceed

$$\frac{1.2\,g/L}{58.5\,g/gmol} = 0.02\ gmol\,/\,L$$

Since we plan to neutralize acid which has an average concentration of 0.014 *gmol/L*, and the neutralization reaction will produce one mole of salt per mole of acid, the average salt concentration will never be greater than 0.014 *gmol/L*. In addition, the dilution provided by adding the NaOH solution will further decrease the resultant salt concentration, so that the average salt concentration will be well within the limits of the state regulations.

There is obviously much more detail needed for the design of the acid-neutralization process. The remainder of this book will be devoted to the further planning, analysis, and evaluation of this process. This effort will require that we learn and apply some fundamental principles of chemical engineering.

At this point, we have a concept for the process we want to develop, so it would be appropriate to draw a simple block diagram or process flow diagram as described in Chapter 2. Let's assume that the HCl waste product can be collected from the manufacturing process and held in a tank that we can use to feed our neutralization process. This is a smart thing to do, because the rate of HCl production may not exactly match the rate of utilization in the neutralization process at every minute; the tank will allow these two processes to be operated independently. Let's also assume that we will have our NaOH available in another tank to use in the neutralization process. Finally, we will assume that a reactor will be necessary for the neutralization reaction to take place. Thus, the process flow diagram (without a stream table) would look something like Figure 3.1.

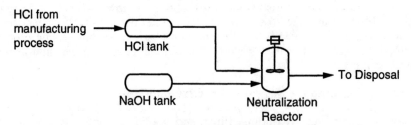

Figure 3.1 Process flow diagram for acid neutralization concept

The Use of Teams in Solving Problems

Up to now in this chapter we have discussed several aspects of problem solving which should be considered by chemical engineers as they solve problems. However, today's engineer does not work in a corner by himself/herself. Rather, most problem solving is done by teams of engineers working together. These teams are often multidisciplinary, bringing together individuals with a variety of different experiences and training. A key assumption is that the team working together can accomplish more than the sum of the efforts of each individual team member working alone. Unfortunately, not all teams work effectively. Our objective in this section is to provide a very brief introduction to several aspects of teamwork which help teams to function more effectively.

Ingredients for a Successful Team

Ingredients which contribute to a successful team include the following[1]:
1. A clear mission or set of goals
2. A plan for attacking problems
3. Clearly defined roles
4. Clear communication
5. Well-defined decision procedures
6. Balanced participation
7. Established ground rules
8. Awareness of group processes

Whole courses can be given on each of the ingredients listed. Team projects should be planned with these elements in mind, and thought should be given to how to incorporate them into the team operation.

Learning to Work Together

Even with important team ingredients (as listed above) in place, assigning a group of people to work together does not make them an effective team. In order to work together effectively, the members of the group must learn to accept each other and to utilize their respective talents for the benefit of the group. How does this work in practice? Researchers have been able to define several distinct developmental stages related to groups[1]. The stages are summarized in Table 3.1.

Table 3.1 Stages of Team Development (adapted from ref. 1)

1.	Forming	Organization of the group, setting of rules and procedures, introductions of members and learning a little about each other.
2.	Storming	Emergence of conflict caused by different perspectives, experiences, backgrounds and views. This is the time when most groups will fail.
3.	Conforming	Coming to the agreement to disagree; tolerance of varying views and opinions and perspectives. Individuals accept the team, their roles on the team, and the individuality of the various team members.
4.	Performing	Utilization of individual differences for the benefit of the group and the work of the group. Varying perspectives and differences are viewed as advantages rather than hindrances.

The performing stage at the bottom of Table 3.1 is the most productive stage, so the goal is to move a team through the other developmental stages to the performing mode as quickly as possible. Some teams are able to reach the performing stage in just a couple of meetings while others may take weeks or even months to reach the same level of performance. Recognition of the stages in group development helps to smooth the transition to the performing stage.

Diversity

One of the real advantages of a team is that it brings people of different talents, abilities and experience together to work on a common objective. It is this diversity that enables a team to be more than just a sum of the individuals who make up the team. This diversity can also lead to conflict (Storming, Table 3.1) as the team members learn to work together. In order to minimize conflict and fully utilize the potential of the team, each member of the team should have a clearly-defined role. Ideally this role should correlate with the strengths of each individual member of the team. To describe such strengths, the responses of people to a goal or task have been classified into four general categories[2]:

Table 3.2 Characteristic Responses to a Goal or Task (adapted from ref. 2)

Fact Finder	Precise, judicious and thorough, this mode deals with detail and complexity, seeking to be both objective and appropriate. Keen at observing and at gathering information, sometimes Fact Finders can be too judicious, seeming overly cautious as they wait for more data. *Keyword: probe*
Follow Thru	Methodical and systematic, this mode is focused and structured, and brings order and efficiency. Follow Thru people are meticulous at planning, programming, and designing, and predictability is essential to their being. *Keyword: pattern*
Quick Start	With an affinity toward risk, this mode is spontaneous and intuitive, flexible, and fluent with ideas. Quick Starters are deadline- and crisis- oriented. They need an atmosphere of challenge and change, and sometimes they can be impatient. *Keyword: innovate*
Implementer	Hands-on, craft-oriented, this mode brings tangible quality to actions. Implementers have a strong sense of three-dimensional form and substance and the ability to deal with the concrete. *Keyword: demonstrate*

Different individuals have different levels of preference for each of these modes of action. Some people have a strong preference for a single mode (e.g., Quick Start) which dominates their goal-driven actions. An individual may also strongly resist action in a particular mode or modes. Others are able to accommodate all four modes of action, adapting to whatever mode is necessary. This last group of people are referred to as facilitators. Facilitators can play a very important role in team work since they are able to work with and accommodate individuals with various action-mode preferences.

There are several important points which can now be made. First, people are different. Not everyone will respond to a task in the same fashion. Understanding this fact is critical to your success in working on teams. Second, there is not one "right" or correct mode of action. The different modes of action represent different talents and/or preferences. A team provides the opportunity for these different talents to be used together to accomplish a shared objective.

Third, an attempt should be made to match an individual's talents with his/her role on the team. It is not very productive to send someone who strongly prefers the Quick Start mode on a fact-finding mission. Similarly, an individual who acts dominantly in the Fact Finder mode cannot be expected to make spontaneous intuitive decisions. Ideally, teams should be structured to take full advantage of everyone's talents. Examples of roles that team members may have within that structure are outlined in Table 3.3. Other tasks, such as preparation of presentations and reports, charting data, etc., are assigned to members of the team as needed.

Table 3.3 Roles Within a Team Structure

Leader/Coordinator	The leader is responsible for calling group meetings, handling or assigning administrative details, planning team activities, and overseeing preparation of reports and presentations.
Observer/Summarizer	This individual is responsible for observing the operation of the group and summarizing key issues.
Data Gatherer	This individual or group of individuals is responsible for gathering data needed for the team to accomplish its goals. Data gathering is typically accomplished between team meetings. It may take the form of gathering quantitative data or may consist of qualitative observations, etc.
Devil's Advocate	Having a "devil's advocate" on the team is useful in probing and evaluating the work of the team. Formal recognition and use of this role turns what might be perceived as a negative contribution into a positive and important part of the total group effort.
Recorder	The recorder writes down the groups decisions and edits the group's report.

One of the important messages of this discussion is that engineering involves the collective contributions of teams, and that those contributions are affected by the differences between people and their ability to mold their team into a cooperative unit. It is useful to illustrate how such differences and team dynamics might work. Let's imagine that you were part of a team that was assigned to address the acid-disposal problem and that your team arrived at the decision to add base to our acid as described earlier in this chapter. The following is a scenario of how that interaction might have taken place:

Meeting #1: Your team of four engineers (with you appointed as team leader) was called together to begin forging a solution to the problem of the acid waste. You took some time in the first meeting to have each team member introduce himself/herself before getting started. You also summarized the problem and the time frame allotted by management for completion of the project. Having defined the problem, you then initiated a brain-storming session to generate a list of possible solutions. One of the team members, Allyson (a relatively new engineer hired about 3 years ago) was particularly good at generating ideas. In contrast, Peter (an experienced engineering nearing retirement) was openly negative about many of the suggestions and kept trying to close the discussion so that analysis could begin on the suggestions that were already on the board. You found yourself getting a little irritated with his disruptive influence. You also wondered if he was one of those engineers who always wanted to analyze everything to death and never seemed to get anything done. Finally, you felt satisfied with the list of possible solutions and you asked the group how they would like to proceed. Lee (a mid-career engineer and meticulous planner) suggested that we meet together the next day to perform a preliminary evaluation of the ideas. We would then be in a better position to define specific tasks and set a regular meeting time. Everyone agreed and the meeting was adjourned.

Meeting #2: The preliminary evaluation began well with evidence that both Allyson and Lee had spent some time preparing for the meeting. However, Peter was openly critical of everyone and everything. He also boasted of his experience and even began to belittle Allyson in an attempt to be humorous. She finally told him to "shut his mouth," and the conversation between the two became heated. It was clear that the meeting was going nowhere so you quickly adjourned before things got out of control.

Your Office: After allowing things to cool off overnight, you called Peter into your office. You assured him that his experience and abilities were important to the success of the team. However, his negative comments and criticism were destroying the ability of the group to work together. Therefore, it was important that such behavior stop immediately. You reviewed the stages of team development with him and pointed out that the team was now at a critical stage. In a very real sense, the success of the team depended on his willingness to make a positive contribution. He assured you that he would make the necessary changes. You also encouraged him to apologize to Allyson.

You also met briefly with Allyson, thanking her for her important contributions to the group and requesting her continued efforts and patience.

Meeting #3: The team met again to complete the preliminary analysis of the possible solutions. Peter made a conscious effort to control his comments and even to be supportive. Allyson recognized his efforts and requested his opinion on issues several times during the meeting. The team was able to complete a preliminary ranking of the ideas, and each member of the group was assigned to perform a more detailed analysis of one of the top options.

Meeting #4: Results were presented from the analyses performed since the last meeting. Respect was demonstrated as team members asked questions, expressed opinions, and/or offered suggestions. Advantages and disadvantages were listed for each alternative, as well as the potential cost and feasibility of completing the work within the specified time frame. Peter was particularly helpful in assessing technical feasibility. Lee's experience with planning and costing was also extremely valuable. Finally, you concluded that it was time to make a decision and you recommended that the team pursue the plan to neutralize the acid with sodium hydroxide. The others agreed. Lee was asked to take the lead in developing specific tasks and a timeline for completion of the project. A regular meeting time was set and the meeting was adjourned.

Summary

This chapter began with a discussion of the steps involved in problem solving. Following those steps, we defined the problem as needing to deal with the acid waste. Then we listed some possible solutions, including changing the company process, finding another contractor to dispose of the waste, building large tanks to store the acid, building an evaporation pond, and treating the acid to prepare it for disposal in the lake. After brief evaluation of the options, the decision was made to pursue the neutralization of the acid to prepare it for disposal in the lake.

Most of this book will center around developing a process to accomplish that goal, and additional chemical engineering principles will be introduced as they are needed for that development.

In today's world, the problem-solving steps mentioned above would most likely be undertaken by teams of engineers. The advantage of teams is that they bring together people of different talents, abilities and experience to work on a common problem. Because of those differences, the team can accomplish much more than what would be accomplished by the team members working alone. However, it takes effort to make a team work effectively, and a knowledge of issues related to teamwork is useful in helping teams to reach their full potential.

References

1. Scholtes, P.R., *The Team Handbook for Educators*, Madison, WI: Joiner Associates Inc., 1994.

2. Kolbe, K., *The Conative Connection*, Menlo Park, CA: Addison-Wesley Publishing Co., 1990.

READING QUESTIONS:

1. Of the seven steps suggested in this chapter for solving engineering problems,

    a. which steps involve thinking of new plans or ideas?

    b. which steps involve evaluating the plans or ideas?

2. In eliminating the possibility of contracting with an independent company to take the acid away, what assumptions were made?

3. In evaluating the possibility of an evaporation pond to take care of the acid, several technical variables were mentioned that would affect the feasibility of such a pond. Based on those variables and any others you think are important, how feasible would an evaporation pond be if the facility were located in your home town? Explain your answer.

4. What two properties of the waste stream would need to be altered before the stream could be discharged to the lake?

5. Why was it necessary to check the state limit on dissolved solids in waste streams in order to determine if the plan to neutralize the acid was a feasible one?

6. What is the key assumption behind the use of teams? Under what conditions do you think this assumption would prove to not be true?

7. Which role in Table 3.3 do you think that you would best play in a team?

8. What benefit(s) can come from knowing the stages of team development (Table 3.1)?

HOMEWORK PROBLEMS:

1. Suggest another possible solution to the problem of dealing with the waste acid and discuss some possible advantages and disadvantages of your proposed solution.

2. Using a computer word processor of your choice, write a memo to your boss in which you communicate...
   a. that you received her memo assigning the problem to you
   b. that you have met with your team to review possible ideas
   c. some of the ideas that had the most merit and their apparent merits and shortcomings (<u>briefly</u> summarize the issues discussed in the book)
   d. your preliminary conclusion about the strategy to pursue
   e. that you will proceed to formulate a preliminary design
   f. that you will advise her of progress on the project and, at the appropriate time, will bring the design to her for input from her and from other important teams in the division

3. As an engineer working for a company which produces food products, you discover a byproduct from one of the company processes. The byproduct is mostly protein (it could be edible) and has the unusual ability to absorb large amounts of water so that it can swell up to 20 times its original volume. As a brainstorming exercise, list 10 possible uses for this product that might become a basis for producing and marketing it.

4. As teams (organized in class or outside of class, depending on your instructor's preference), discuss each of the following (each student should turn in a separate set of answers): Considering each personality type described in Table 3.2...

   a. Which role(s) described in Table 3.3 would each personality type tend to play?
   b. How would each personality type tend to contribute to the problem-solving steps outlined at the beginning of this chapter?

5. Do the following problem as teams (organized in class or outside of class, depending on your instructor's preference):

   Suppose that you work for a large power company. One of your company's holdings is a small coal-fired power plant near an old coal mine which provided fuel for the plant. The plant is not currently in operation due to a long history of fires and cave-ins which have caused a number of deaths and led to closure of the mine. However, a modern (and safe) coal mine is in operation approximately 100 miles away. Existing railroad tracks cover only 30 miles of the distance between the plant and the operational mine.

   a. Based on the above information, propose at least three possible problem definitions. Which do you feel is most appropriate?

   b. Propose 4 possible strategies to put the power plant in operation.

   c. For each strategy, indicate what information would be needed in order to evaluate the merits of each strategy. You should include 3-5 <u>information items</u> per strategy.

   d. Suppose that you are a manager employed by the power company, and that you have been given responsibility for the problem described above. Organize a hypothetical group of five people to address this problem and propose a plan for its solution. Specify the roles that these individuals might fulfill as part of the group. What types of individuals (see Table 3.2) would be best suited to these roles?

6. After the team activity of Problem 5, answer the following individually:

   a. Describe the team interactions you observed as your team worked on Problem 5. Particularly, explain whether you identified elements of team development as outlined in Table 3.1.

   b. From the descriptions in Table 3.2 and from lifelong experience with yourself, identify which type of responder you tend to be. Did you function that way during your brief team effort with Problem 5?

   c. What type of responder (from Table 3.2) did each of your team members appear to be?

Courtesy of Smoot Company, Kansas City, KA
and Jewell Baker Zander - Chemical Advertising Specialists, Kansas City, MO

CHAPTER 4

DESCRIBING PHYSICAL QUANTITIES

As we begin to design the acid-neutralization process (or any other process, for that matter), it will be important to describe the various process variables. This will require an understanding of how to express the values of physical quantities. In order to describe some of these quantities, we will rely on a few of the principles and concepts from chemistry, including molecular weight, moles, and concentration. You are encouraged to refer to a high-school or college chemistry text for review or for additional background if needed.

Section 4.1   Units

All physical quantities have a *numerical value* (e.g. 6.51) and a *unit* (indicating what there are 6.51 of, such as feet, grams, seconds, etc.). For example, in our problem, we will want to assess how well we have neutralized the acid; thus, we will need to talk about the acid "concentration." To do that, we will need to know how *concentration* is defined, and we will need to recognize the various ways that concentration can be expressed. A description of other process variables will also be needed. This chapter will review some of the more important variables and units.

Two systems of units in use in most of the industrialized world are the *metric system* and the *American engineering system*. In the metric system, smaller divisions of a base unit measurement and larger combinations of that base unit are related to that original base unit by powers of ten. For example, the measure of length can be a meter, a decimeter (0.1 meters), a centimeter (0.01 meters), a millimeter (0.001 meters), a kilometer (1000 meters), etc. On the other hand, the American engineering system is based on cultural definitions of various measurements from British history (e.g. a yard being the length from the king's nose to the tip of his middle finger on his fully-extended right arm), and relationships between smaller and larger units of the same measure vary widely. For example, there are 12 inches per foot, 3 feet per yard, and 5,280 feet per mile.

Multiple measurement systems have created problems as many products are traded internationally, and parts for many products are manufactured in more than one country or region of the world. In 1960, an international forum of technical experts representing the industrialized world met to deal with this problem and recommended the adoption of a standardized set of units within the metric system. This set, called the *Systeme Internationale d'Unites*, or SI system, has been the standard measurement system toward which many of the industrialized countries, including the United States, have slowly moved for eventual adoption.

In our physical world, we measure and talk about a number of different physical parameters, but a small number of these parameters are the building blocks for all others. For example, the most basic of those parameters or building blocks are length, mass, time, and temperature. We sometimes refer to these as the *basic dimensions* of our world. Other parameters which we use are combinations of these dimensions, as will be discussed later in this section.

Each of the basic dimensions can be expressed in a variety of units (e.g. time can be represented in seconds, minutes, hours, days, etc.). However, one can define subsystems where one unit of measurement is defined as the "base" unit of measurement with other units as multiples of that base unit. For example, the *SI system* mentioned previously is a subsystem of the metric system and was defined with the base unit of length being the meter (*m*), the base unit of mass being the kilogram (*kg*), and the base unit of time being the second (*s*). This is in contrast with another subset of the metric system, the *cgs system*, which uses the centimeter (*cm*), gram (*g*), and second (*s*) as the base units for length, mass, and time, respectively. Table 4.1 summarizes the base units of mass, length, time, and temperature for the cgs and SI subsystems. While the American engineering system is a general system of units (like the metric system) and no base units are strictly defined, sample units for each category have been included in Table 4.1 for comparison.

Table 4.1 Base or Sample Units for Three Measurement Systems

System	mass	length	time	temperature
cgs	*g*	*cm*	*s*	Celsius
SI	*kg*	*m*	*s*	Kelvin
American	*lb$_m$*	*ft*	*s*	Fahrenheit

Conversion Factors

It is frequently necessary to convert from one type of unit to another, for example from inches to feet, from grams to pounds-mass (*lb$_m$*), or from seconds to hours. The key to such conversions is the use of *conversion factors*, defined as follows:

A conversion factor is an equation which equates two quantities expressed in different units where, in fact, the two quantities are exactly equivalent.

The following are a few common conversion factors:

$$12 \ in = 1 \ ft \qquad 1000 \ g = 1 \ kg \qquad 60 \ s = 1 \ min$$

A more complete list of conversion factors is found in the beginning of this book. Each such relationship can also be rearranged into the form of a ratio as follows:

$$1 = \frac{12 \ in}{1 \ ft} \qquad 1 = \frac{1000 \ g}{1 \ kg} \qquad 1 = \frac{60 \ s}{1 \ min}$$

The ratios on the right side of each equation are useful tools for converting from one unit or set of units to another. Such a conversion is accomplished by multiplying the original number (with its units) by the appropriate conversion ratio where the ratio is written as "new unit/old unit." Such multiplication is legitimate because the actual value of each such ratio is "1," as indicated above. To illustrate, a conversion from 28 *inches* to its equivalent number of *feet* would be accomplished as follows:

$$(28 \ in)\left(\frac{1 \ ft}{12 \ in}\right) = 2.333 \ ft$$

Note that the old units are canceled out by corresponding units in the conversion factor, leaving only the new units. That's how you know that you have formulated the conversion factor correctly for this conversion. If the conversion factor had been inverted, for example, the units would have not canceled out properly, giving

$$(28\ in)\left(\frac{12\ in}{1\ ft}\right) \ = \ 336\ in^2/ft$$

which is obviously not in the desired units. Another way to write the conversion equation is to use vertical and horizontal lines instead of parentheses.

$$\frac{28\ in\ \mid\ 1\ ft}{\mid\ 12\ in} \ = \ 2.333\ ft$$

Finally, many conversions will require the use of more than one conversion factor, such as the following conversion of 37759 inches to its equivalent in kilometers (*km*):

$$(37759\ in)\left(\frac{2.54\ cm}{1\ in}\right)\left(\frac{1\ m}{100\ cm}\right)\left(\frac{1\ km}{1000\ m}\right) \ = \ 0.959\ km$$

Moles

In your chemistry classes, you learned about a *gram-mole* of a substance, often just called a *mole*. A gram-mole of a chemical compound is defined as the amount (number of molecules) of that compound whose mass in grams is numerically equal to its *molecular weight* (molecular weight is discussed in greater detail below). A gram-mole is a convenient quantity to use when working in the cgs system. There are two other types of moles as well, namely a *kilogram-mole* and *pound-mole*, which are convenient when working in the SI and American systems, respectively. Like the gram-mole, the kilogram-mole and pound-mole are defined in terms of the *molecular weight* of the substance and represent a certain number of molecules. Following are the definitions of these three types of moles, along with the abbreviated designation for each (i.e. *gmol*, *kgmol*, and *lbmol*):

gram-mole (*gmol*): the amount of a species such that its mass in <u>grams</u> numerically equals its molecular weight (this amount is associated with Avogadro's number of molecules)

kilogram-mole (*kgmol*): the amount of a species such that its mass in <u>kilograms</u> numerically equals its molecular weight

pound-mole (*lbmol*): the amount of a species such that its mass in <u>pounds-mass</u> numerically equals its molecular weight

Molecular Weight

The molecular weight of a molecule is the sum of the masses of all the atoms which make up that molecule. Therefore, molecules which consist of heavy atoms and/or large numbers of atoms will have a higher molecular weight than molecules made up of a few light atoms. In practice, it is inconvenient to measure and use the absolute mass of single molecules expressed in atomic mass units. Instead, we use the mass (e.g., in g, kg, or lb_m) of a large number of molecules (*gmol*, *kgmol*, or *lbmol*). As you learned in your chemistry courses, the number of molecules in a gram-mole is Avogadro's number. It can be shown that the number of molecules in a kg-mole is 1000 times Avogadro's number and that the number of molecules in a pound-mole is 453.6 times Avogadro's number. From the above definitions, the molecular weight of a substance is 1) the mass in grams of one gram-mole of that substance, 2) the mass in kilograms of one kg-mole of that substance, and 3) the mass in pounds-mass of one pound-mole of that substance. For example, for oxygen, the molecular weight is 32 atomic mass units, so 1 *gmol* of oxygen has a mass of 32 g, 1 *kgmol* has a mass of 32 kg, and 1 *lbmol* has a mass of 32 lb_m. Thus, the molecular weight of oxygen is a type of conversion factor and can be written

$$MW_{O_2} = \frac{32\ g\ O_2}{1\ gmol} = \frac{32\ kg\ O_2}{1\ kgmol} = \frac{32\ lb_m\ O_2}{1\ lbmol}$$

Additionally, from the above equation, it is evident that the relationships between the different kinds of moles are similar to the relationships between the corresponding mass units. For example, 1 *kgmol* = 1000 *gmol*, and 1 *lbmol* = 454 *gmol*.

Symbols

In the quantitative descriptions and calculations used in chemical engineering, it is frequently necessary to represent physical quantities using symbols. We will assign symbols to physical parameters throughout this book and now introduce the following symbols to represent some of the quantities we have discussed thus far:

m = the mass of a quantity of material
m_A = the mass of a particular chemical species (in this case, species "A"), either as a pure
         material or within a mixture
n = the number of moles of a material
n_A = the number of moles of a particular chemical species (in this case, species "A"),
         either as a pure material or within a mixture
MW_A = the molecular weight of a particular chemical species (in this case, species "A")

In the above symbols which refer to a particular chemical species (m_A, n_A, and MW_A), the identity of that species is indicated in the subscript. For any particular compound, say hydrochloric acid, the subscript "A" would be changed to indicate that compound, i.e. m_{HCl}.

In terms of the symbols just introduced, the molecular weight of a chemical compound (e.g. species "A") can be written

$$MW_A = \frac{m_A}{n_A} \tag{4.1}$$

The use of these symbols and the molecular weight to perform conversions between mole and mass units is illustrated in Example 4.1.

Example 4.1

Common table sugar is sucrose, $C_{12}H_{24}O_{12}$. How many *lbmol* of sucrose are in a bag that has a mass of 100 lb_m? How many *kgmol*?

Solution: Using atomic weights from the list in the front of the book, the molecular weight of sucrose ($MW_{sucrose}$) is   12(12)+24(1)+12(16)=360. The mass of 100 lb_m can also be expressed in kilograms using the following conversion

$$\text{Mass in } kg:\ m_{sucrose} = (100\ lb_m)\left(\frac{0.45359\ kg}{1\ lb_m}\right) = 45.4\ kg$$

Rearranging Equation 4.1

$$\text{Determine } lbmol:\ n_{sucrose} = \frac{m_{sucrose}}{MW_{sucrose}} = \frac{100\ lb_m}{360\ lb_m/lbmol} = 0.278\ lbmol$$

$$\text{Determine } kgmol:\ n_{sucrose} = \frac{m_{sucrose}}{MW_{sucrose}} = \frac{45.4\ kg}{360\ kg/kgmol} = 0.126\ kgmol$$

Derived Units

Many units are *derived units*, i.e. are derived as combinations of the base units. For example, units of velocity are derived from length and time, hence "*mi/hr*." Examples of a number of derived units are listed in Table 4.2 for the three measurement systems discussed above. A number of these derived units will be discussed later in this chapter.

Table 4.2 Examples of Derived Units for Three Measurement Systems

System	cgs system	SI System	American System
density	g/cm^3	kg/m^3	lb_m/ft^3
velocity	cm/s	m/s	ft/s
acceleration	cm/s^2	m/s^2	ft/s^2
volumetric flow rate	cm^3/s	m^3/s	ft^3/s
mass flow rate	g/s	kg/s	lb_m/s
concentration	$gmol/cm^3$	$kgmol/m^3$	$lbmol/ft^3$

To produce conversion factors for derived units, conversion factors of base units can be raised to any power. The following are a few examples:

$$\text{area:} \qquad \left(\frac{100 \; cm}{1 \; m}\right)^2 = \frac{10^4 \; cm^2}{1 \; m^2}$$

$$\text{volume:} \qquad \left(\frac{12 \; in}{1 \; ft}\right)^3 = \frac{1728 \; in^3}{1 \; ft^3}$$

Notice that raising a conversion factor to a certain power raises its units to that same power.

Force

An important parameter which has derived units is force, and one kind of force is weight. The force "F" to accelerate a mass "m" at an acceleration rate "a" is defined by Newton's second law as

$$F = ma \tag{4.2}$$

One form of acceleration is that associated with earth's gravity. (This is an acceleration because holding all matter on the earth's surface requires constantly changing the direction of travel to produce a circular arc.) In this case, the rate of acceleration is described by the gravitational acceleration, g, and the force that an object exerts on the earth's surface (its "weight"), is

$$W = mg \tag{4.3}$$

In the American engineering system, g is expressed in ft/s^2, so the weight of a person whose mass was measured in lb_m would be in units of "$lb_m \, ft/s^2$." These units are not very convenient, so an *equivalent unit* of pound-force (lb_f) has been defined as 1 $lb_f \equiv 32.174 \; lb_m \, ft/s^2$ (the "triple equals" sign "$\equiv$" denotes a definition). In this way, weight can be calculated and expressed in the convenient unit of lb_f (usually simply called "pounds"). Table 4.3 summarizes the values of g (at sea level) and the equivalent units of force for the three units systems discussed previously.

Table 4.3 Gravitational Acceleration (at sea level) and Equivalent Units of Force

System	g	Equivalent Unit of Force
cgs	980.66 cm/s²	1 dyne ≡ 1 g cm/s²
SI	9.8066 m/s²	1 Newton (N) ≡ 1 kg m/s²
American	32.174 ft/s²	1 pound-force (lbf) ≡ 32.174 lbm ft/s²

Each of the definitions of force units (dyne, Newton, and pound-force) can also be rearranged into the ratio form of a conversion factor as follows:

$$1 = \frac{1\ g\ cm}{s^2 dyne} = \frac{1\ kg\ m}{s^2 N} = \frac{32.174\ lb_m ft}{s^2 lb_f}$$

This conversion factor is used so frequently, it is sometimes referred to with the symbol g_c. Of course, like all conversion factors, the inverse of each expression is also a conversion factor:

$$1 = \frac{1\ s^2 dyne}{g\ cm} = \frac{1\ s^2 N}{kg\ m} = \frac{1\ s^2 lb_f}{32.174\ lb_m ft}$$

The use of such conversion factors is illustrated in Examples 4.2 and 4.3.

Example 4.2

An object has a mass equal to 1 lb_m. What is its weight in pounds-force (lb_f)?

Solution:

From Table 4.3, we note that $g = 32.174\ ft/s^2$. From Equation 4.2,

$$W = mg = (1\ lb_m)\left(\frac{32.174\ ft}{s^2}\right)\left(\frac{1\ lb_f s^2}{32.174\ lb_m ft}\right) = 1\ lb_f$$

This result explains why a pound-force was defined as it was — so that one pound-mass would weigh one pound-force (on the earth). However, you should understand that <u>it is never correct to simply write that one pound-force *equals* one pound-mass!</u>

Example 4.3

An object has a mass equal to 8.41 kg. What is its weight a) in Newtons (N) and b) in pounds-force (lb_f)?

Solution:

a. From Table 4.3, we note that $g = 9.8066 \ m/s^2$. Thus, from Equation 4.2,

$$W = mg = (8.41 \ kg)\left(\frac{9.8066 \ m}{s^2}\right) = 82.5 \ N$$

b. From the answer to part a,

$$W = (82.5 \ N)\left(\frac{0.22481 \ lb_f}{1 \ N}\right) = 18.5 \ lb_f$$

Now, we are ready to discuss some of the variables that will be important to the problem we are trying to solve.

Section 4.2   Some Important Process Variables

As we design the process for neutralizing the HCl, we will need to describe the flow rate of material into and out of that process. For that description, we need to understand how to express fluid density, flow rate, and chemical composition.

Density

The *density* of a material is the <u>mass</u> of a <u>unit volume</u> of that material. A common symbol to represent density is the Greek letter ρ (pronounced "rho"). For most liquids and solids, the densities are listed in standard reference books, such as the <u>CRC Handbook of Chemistry and Physics</u>[1] or <u>Perry's Chemical Engineering Handbook</u>[2]. For liquids, the density is relatively independent of such variables as pressure and temperature. A handy density to remember is the approximate density of water at room temperature, which is

$$\rho_{water,25°C} \approx 1.0 \ g/cm^3 = 1000 \ kg/m^3$$

As we would expect, the density of gases are much smaller than those of liquids. For comparison, the density of air at room temperature is

$$\rho_{air,25°C} \approx 0.0012 \ g/cm^3 = 1.2 \ kg/m^3$$

For gases, the density varies significantly with the temperature and pressure of the gas, as you may have learned in your chemistry classes. Typical units for density are listed in Table 4.2.

The density of a material can be used to relate the volume (V) of a material to its mass. For example, the density of liquid propane is $36.53 \ lb_m/ft^3$. Therefore, the mass of 500 *gal* would be

$$m = \rho V = \left(\frac{36.53 \ lb_m}{ft^3}\right)(500 \, gal)\left(\frac{1 \, ft^3}{7.4805 \ gal}\right) = 2442 \ lb_m$$

Flow Rate

We often describe the rate at which the stream flows using the *flow rate* (the amount of material which passes a reference point within a unit time). Three common types of flow rates used are *mass flow rate* (symbol: $\dot{m}$): the mass of a material which passes a reference plane within a unit time interval

molar flow rate (symbol: $\dot{n}$): the number of moles of a material which passes a reference plane within a unit time interval

volumetric flow rate (symbol: $\dot{V}$): the volume of a material which passes a reference plane within a unit time interval

With the symbols assigned, the "dot" over the letter indicates that the variable is a "rate" (i.e. a measure of change per time). Typical units for volumetric and mass flow rate are listed in Table 4.2. To better understand these flow rates, it might be helpful to visualize water coming out of a kitchen faucet, where the mouth of the faucet could represent the reference plane in the above definitions. For example, if the volumetric flow rate of water from the faucet is 2.5 *gal/min*, we could collect the water coming from the faucet for one minute, and we would find that we had collected 2.5 gallons. If we collected for two minutes, we would find that we had collected 5 gallons, etc.

Conversion between mass flow rate (mass/time), volumetric flow rate (volume/time), and density (mass/volume) is

$$\bigstar \quad \dot{m} = \rho \dot{V} \tag{4.4}$$

You should understand the concepts and units in this equation and <u>be able to reproduce this equation without looking at the book</u>. Now let's apply these variables to the acid-neutralization problem.

Example 4.4

The average flow rate of HCl produced by our company is 11,600 *L/hr* (see Chapter 3). It's density is approximately the same as that of water, namely 1000 kg/m^3 or 1 kg/L.

a. What is the equivalent of this flow rate in units of cm^3/s?

b. What is the mass flow rate for this stream in kg/hr?

Solution:

a. Applying the appropriate conversion factors

$$\dot{V} = \left(11,600\frac{L}{hr}\right)\left(\frac{1000\ cm^3}{1\ L}\right)\left(\frac{1\ hr}{60\ min}\right)\left(\frac{1\ min}{60\ s}\right) = 3,222\frac{cm^3}{s}$$

Again, notice that the units canceled out to leave the units we were seeking, indicating that we used the conversion factors correctly.

b. From Equation 4.4

$$\dot{m} = \rho\dot{V} = \left(\frac{1000\ kg}{1\ m^3}\right)\left(\frac{11,600\ L}{1\ hr}\right)\left(\frac{1\ m^3}{1000\ L}\right) = 11,600\frac{kg}{hr}$$

Mixture Composition

In dealing with mixtures, substances which contain more than one chemical compound or specie, it is often important to describe the *composition* of the mixture. In other words, we often wish to know how much of each individual compound there is within the mixture. The most common expression of composition is *concentration*, which is the number of moles of the particular species per volume of mixture. The concentration of species A is usually represented by the symbol c_A, thus

$$\text{Concentration of A: } c_A = \frac{\textit{moles of A}}{\textit{volume of mixture}} = \frac{n_A}{V} = \frac{\textit{molar flow rate of A}}{\textit{volumetric flow rate of mixture}} = \frac{\dot{n}_A}{\dot{V}}$$

One form of concentration is *molarity* (abbreviated with the letter "*M*"), which is defined as the number of *gmol* per *liter* of mixture or solution. Other units of concentration can be used as well, and some are listed in Table 4.2.

Besides concentration, several additional methods are used to define the composition of a mixture of substances where one of those substances is species A. They are

$$\text{Mass Fraction of A: } x_A = \frac{\textit{mass of A}}{\textit{mass of mixture}} = \frac{m_A}{m} = \frac{\textit{mass flow rate of A}}{\textit{mass flow rate of mixture}} = \frac{\dot{m}_A}{\dot{m}}$$

$$\text{Mole Fraction of A: } y_A = \frac{\textit{moles of A}}{\textit{moles of mixture}} = \frac{n_A}{n} = \frac{\textit{molar flow rate of A}}{\textit{molar flow rate of mixture}} = \frac{\dot{n}_A}{\dot{n}}$$

Mass Percent of A = 100 x_A (commonly expressed as weight percent and abbreviated *wt%*)

Mole Percent of A = 100 y_A (abbreviated *mole%*)

In the calculation of the mass fraction and mole fraction, the same units for mass and moles must be used in the numerator as in the denominator so that the calculated fraction is without units.

With these definitions, we can summarize the relationships between them as follows:

$$\dot{m}_A = x_A \dot{m} = MW_A \dot{n}_A = MW_A y_A \dot{n} = MW_A c_A \dot{V} \tag{4.5}$$

As with Equation 4.4, it is important that you fully understand and can reconstruct these relationships. For example, suppose that you know the value of the mass flow rate of a chemical compound (with known molecular weight) within a mixture and that you also know the volumetric flow rate of that mixture. You should be sufficiently familiar with the relations in Equation 4.5 to be able quickly to compute the concentration of the compound in the mixture. Furthermore, you should also be very familiar with the relations and terms in Equation 4.4 so that you can use Equations 4.4 and 4.5 together. Make sure you understand the concepts and units of all variables in both equations and practice writing them until you can reproduce both equations without looking at the book.

Example 4.5

For the acid-neutralization problem, the volumetric flow rate of the HCl solution coming from our manufacturing process is 11,600 *L/hr*, and the average molarity of HCl in that stream is 0.014 *M*. Based on this information and the answers in Example 4.4,

a. How many *gmol* of HCl are in 88 m^3 of the solution?

b. How many *gmol* of HCl are flowing from the process per minute (i.e. what is the molar flow rate) when the volumetric flow rate of the solution is 11,600 *L/hr*?

c. What is the mass fraction of HCl in the solution?

Solution:

a. $n_{HCl} = c_{HCl}V = \left(\dfrac{0.014\ gmol}{L}\right)(88m^3)\left(\dfrac{1000L}{m^3}\right) = 1,232\ gmol\ HCl$

b. $\dot{n}_{HCl} = c_{HCl}\dot{V} = \left(\dfrac{0.014\ gmol}{L}\right)\left(\dfrac{11,600\ L}{hr}\right)\left(\dfrac{1\ hr}{60\ min}\right) = 2.71\ gmol\ HCl/min$

c. $x_{HCl} = \dfrac{\dot{m}_{HCl}}{\dot{m}} = \dfrac{MW_{HCl}\dot{n}_{HCl}}{\dot{m}}$

Atomic weights are listed in the front of the book, giving $MW_{HCl} = 1.0 + 35.5 = 36.5$ g HCl/gmol, and $\dot{m}$ is available from Example 4.4

so $x_{HCl} = \left(\dfrac{(36.5\,g\ HCl/gmol)(2.71\,gmol\ HCl/min)}{11,600\,kg/hr}\right)\left(\dfrac{1\,kg}{1000\,g}\right)\left(\dfrac{60\,min}{1\,hr}\right) = 0.00051$

Conversion Between Mole Fraction and Mass Fraction

It is common for a chemical engineer to know the mass fractions or percentages and to need the mole fractions or percentages or to know the mole fractions or percentages and to need the mass fractions or percentages. In such cases, it is necessary to convert between the various types of fractions or percents. When performing such conversions, the first step is to assume an amount of material for the calculation (called the *basis of calculation*). The steps outlined in Figures 4.1 and 4.2 are suggested. You will note that when the given compositions are expressed as fractions or percentages, the most convenient basis of calculation is 100 units of mass or moles (e.g. 100 g, 100 kgmol, etc.) so that the amount of each component is easily calculated. Example 4.6 illustrates the application of the procedure outlined in Figure 4.1

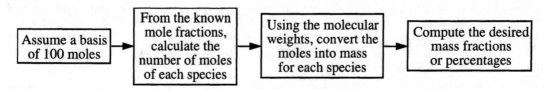

Figure 4.1 Strategy for Converting Mole Fractions /Percentages to Mass Fractions/Percentages

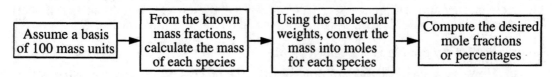

Figure 4.2 Strategy for Converting Mass Fractions/Percentages to Mole Fractions/Percentages

Example 4.6

Air has the following approximate mole percentages:

N_2 78.03 *mole%*
O_2 20.99 *mole%*
Ar   0.94 *mole%*

What are the mass percentages of these components?

<u>Solution</u>: As outlined in Figure 4.1, we'll select a <u>basis of 100 *gmol* of air</u> and calculate how many *gmol* of each substance are present in those 100 *gmol*.

N_2: $n_{N_2} = y_{N_2} n = \left(\dfrac{78.03 \ gmol \ N_2}{100 \ gmol \ total} \right) (100 \ gmol \ total) = 78.03 \ gmol \ N_2$

O_2: $n_{O_2} = y_{O_2} n = \left(\dfrac{20.99 \ gmol \ O_2}{100 \ gmol \ total} \right) (100 \ gmol \ total) = 20.99 \ gmol \ O_2$

Ar: $n_{Ar} = y_{Ar} n = \left(\dfrac{0.94 \ gmol \ Ar}{100 \ gmol \ total} \right) (100 \ gmol \ total) = 0.94 \ gmol \ Ar$

Continuing to follow the strategy in Figure 4.1, we'll now calculate the number of <u>*grams*</u> of each substance.

N_2: $m_{N_2} = MW_{N_2} n_{N_2} = (28 \ g \ N_2/gmol)(78.03 \ gmol \ N_2) = 2185 \ g$

O_2: $m_{O_2} = MW_{O_2} n_{O_2} = (32 \ g \ O_2/gmol)(20.99 \ gmol \ O_2) = 672 \ g$

Ar: $m_{Ar} = MW_{Ar} n_{Ar} = (39.95 \ g \ Ar/gmol)(0.94 \ gmol \ Ar) = 38 \ g$

Finally, the <u>total</u> number of *grams* is 2185 + 672 + 38 = 2895 *g*, so

$$x_{N_2} = \frac{m_{N_2}}{m} = \frac{2185 \ g \ N_2}{2895 \ g} = 0.755$$

$$x_{O_2} = \frac{m_{O_2}}{m} = \frac{672 \ g \ O_2}{2895 \ g} = 0.232$$

$$x_{Ar} = \frac{m_{Ar}}{m} = \frac{38 \ g \ Ar}{2895 \ g} = 0.013$$

By the way, one should always check the calculation to see if the mass fractions add up to 1.000. Check: 0.755 + 0.232 + 0.013 = 1.000!

Dimensional   Consistency

Equations which correctly describe physical phenomena must obey the rules of *dimensional consistency*. Those rules are:

1. Terms which are added together (or subtracted) must have the same units. For example, in the equation $Q = ab + c^2$, the units of ab must be the same as those of c^2.

2. Exponents must be unitless. Thus, if an exponent consists of several terms, the units of all those terms must cancel. For example, in the equation $y = x^{ab/c}$, the units in the term ab/c must all cancel out to leave no units.

Keeping track of the units in a calculation provides a number of benefits. One that has already been mentioned is that it provides a way of safeguarding against the incorrect application of conversion factors. It also provides a check on the equation being used for the calculation, because an error in that equation might produce erroneous units for the answer (e.g. "*kg m/s²*" for a velocity!). An error in the equation might also produce a violation of the rules of dimensional consistency, which would be discovered as units are carefully monitored. For these reasons, the habit of giving careful attention to units is vital to good engineering practice, and the reader should work hard to establish that habit in all technical calculations.

References:

1. *Handbook of Chemistry and Physics*, D.R. Lide, ed., 71st ed., Cleveland, OH: CRC Press, Inc., 1990.

2. *Perry's Chemical Engineering Handbook*, R.H. Perry and D.W. Green, eds., 6th ed., NY: McGraw-Hill Book Co., 1984.

READING QUESTIONS:

1. The definition of "pressure" will be presented in Chapter 7 as force per area. In a particular situation, the pressure exerted on a certain surface is found to be 94,000 *N/m²* Identify the following parts of this expression of pressure:

    > the numerical value
    > all basic dimensions represented
    > all base units involved
    > any derived unit(s)

2. In this chapter, we mention that problems arise from having at least two systems of units in the industrialized world (i.e. the metric system and the American engineering system). In more specific terms, describe at least two such problems that you think would arise?

3. The length of a specimen is determined and, in the cgs system, is represented as 3.2 *cm*. How would this same length be represented in the SI system?

4. The weight of an astronaut is measured on a distant planet and found to be one-fifth of his weight on the earth's surface. Is his mass different on that distant planet than on earth? What does the weight difference imply about the acceleration of gravity on the distant planet?

5. In an attempt to compute the number of seconds equivalent to 36 minutes, your colleague obtains for an answer: $0.6\ min^2/s$. What did your colleague do wrong?

6. Using water and air as examples, what is an approximate ratio of the densities of liquids to gases?

7. Two compounds, one with a high molecular weight and one with a low molecular weight, are flowing at the same mass flow rate. Which has the greater molar flow rate?

8. Two compounds, one with a high density and one with a low density, are flowing at the same mass flow rate. Which has the greater volumetric flow rate?

9. Two compounds, one with a high molecular weight and one with a low molecular weight, each comprise 50 mole% of the same mixture. Which has the greater mass fraction in the mixture?

10. Solution 1 has a greater density than does solution 2, and solution 1 also has a greater concentration of species A than does solution 2. For these solutions, answer each of the following questions and support your answer:

    a. Which will occupy greater volume: 1 kg of solution 1 OR 1 kg of solution 2

    b. Which will contain more molecules of species A: 1 gallon of solution 1 OR 1 gallon of solution 2?

    c. If the two solutions flow with equal volumetric flow rate, which stream will have:

       i)   the greater mass flow rate for the entire stream?

       ii) the greater molar flow rate of species A?

       iii) the greater mass flow rate of species A?

11. A solution of salt dissolved in water is diluted with additional water. With each of the following variables, indicate whether the dilution process will cause the value of the variable to increase, decrease, or stay the same. Support your answer.

    a. x_{salt}

    b. V

    c. c_{salt}

    d. m_{salt}

12. A solution of NaOH in water flows in a stream, and the mass flow rate of the stream is suddenly increased. For each of the following properties of the stream, indicate whether the increase in flow rate will cause the property to increase, decrease, or remain the same. In each case, explain your answer.

    a. ρ e. $\dot{n}$

    b. c_{NaOH} f. $\dot{V}$

    c. $\dot{m}_{NaOH}$ g. MW_{NaOH}

    d. y_{NaOH}

HOMEWORK PROBLEMS:

1. Perform the following conversions by determining the equivalent value of the given number in the new units indicated:

   a. 3.9 *cm/s*   to   *mi/hr*

   b. 177 $lb_m\, ft/min^2$ to $kg\, cm/s^2$

   c. 47 ft^3 to   *gal*

2. Water at 4°C has a density of 1000 kg/m^3. What is the equivalent value in units of

   a. g/cm^3 ?

   b. lb_m/ft^3 ?

   c. $gmol/L$? (as used in Example 4.5)

3. A gas mixture has the following percentages by mass:

   $$
   \begin{array}{ll}
   N_2: & 70\% \\
   O_2: & 14\% \\
   CO: & 4\% \\
   CO_2: & 12\%
   \end{array}
   $$

   What are the mole percentages?

4. A 6*M* sulfuric acid solution (H_2SO_4 in water) is flowing into a tank at the rate of 100 *liters/min*. The density of the solution is 1.34 g/cm^3. Please determine the value of the items requested below (show your work). For parts (b)-(d) first write out the equation using symbols for the appropriate variables, then algebraically solve for the unknown variables, and finally substitute in the numerical values and calculate the values for the unknown variables.

   a. The molecular weight of the sulfuric acid

   b. The molar flow rate of H_2SO_4 into the tank

   c. The mass flow rate of H_2SO_4 into the tank

   d. The total mass flow rate into the tank

5. A stream consisting of two organic chemicals, 1) benzene (C_6H_6) and 2) toluene (C_7H_8), enters a separation column. The total mass flow rate of the stream is 10,000 lb_m/hr. The mass percent of benzene in the stream is 40%. Please determine the following:

   a. The mass flow rate of benzene

   b. The mass flow rate of toluene

   c. The molar flow rate of toluene

   d. The total molar flow rate of the stream

   e. The mole fraction of benzene

6. The exhaust gas coming from a coal-burning furnace (*flue gas*) usually contains sulfur, in the form of SO_2, and when the gas is discharged into the atmosphere (which sometimes happens), the SO_2 can eventually react with oxygen and water to form sulfuric acid (H_2SO_4) — hence, acid rain. The reaction is

$$SO_2 + \tfrac{1}{2} O_2 + H_2O \rightarrow H_2SO_4$$

The air around an old power plant has an average composition as follows:

H_2SO_4	0.1 *mole%*
O_2	20.2 *mole%*
N_2	77.9 *mole%*
H_2O	1.8 *mole%*

What is the number of grams of sulfuric acid per ton (2000 lb_m) of this "air"? (Note: the molecular weight of this humid and polluted "air" cannot simply be assumed to be 29.)

7. The following variables have the units indicated

$$x \text{ has units of } g/s$$
$$y \text{ has units of } cm$$
$$z \text{ has units of } g/cm\ s$$
$$a \text{ has units of } g/s$$
$$b \text{ has units of } cm$$

In terms of these variables, which of the following obey the laws of dimensional consistency? (Support your answers)

a. $w = x/y + ab - z$

b. $J = J_o\, e^{b\left[\frac{z}{x-a} + \frac{1}{y}\right]}$

8. A piston is movable (up and down) inside a vertical cylinder as shown below. The pressure beneath the piston is greater than the pressure on top of the piston, and this difference in pressure can support the mass (and weight) of the piston according to the equation

$$(\text{Pressure}_{\text{beneath}} - \text{Pressure}_{\text{top}})\ \text{Area}_{\text{piston cross section}} = (\text{mass of piston})\ g$$

where "g" is the acceleration of gravity and the other pertinent values are shown in the drawing. Note: pressure is force per area, and the abbreviation "*psi*" on the values of the pressure represents lb_f/in^2. For these values, how much mass (lb_m) can the piston have and be supported by these pressures?

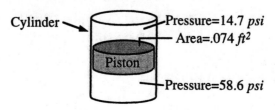

9. When a fluid flows from one location ("start") to an another location ("end") under certain circumstances, the changes in fluid properties can be described by Bernoulli's equation, which is

$$\frac{P_{start} - P_{end}}{\rho} + \frac{1}{2}\alpha(v_{start}^2 - v_{end}^2) + g(z_{start} - z_{end}) = 0$$

where

α = a dimensionless correction factor

ρ = fluid density
P = fluid pressure
v = fluid velocity
z = fluid elevation
g = acceleration of gravity

Prove that this equation is dimensionally consistent in both the American engineering system and the metric system.

Hint: To handle the pressure term, see Problem 8 and also remember the definitions of the various units of force shown in Table 4.3.

CHAPTER 5

STEADY-STATE MATERIAL BALANCES
(HOW MUCH BASE DO WE NEED?)

Section 5.1 Conservation of Total Mass

Now that we have decided to dispose of our acid by neutralizing it with base, and now that we have learned how to describe the physical quantities such as mass and moles of acid and base, we are ready to determine how much NaOH solution would need to be added to the HCl product in order to neutralize it. That determination will require an understanding of *material balances*.

One important principle in dealing with material balances is that <u>total mass is conserved</u>. Ignoring the very small conversion of mass to energy in nuclear reactions, mass will never be created or destroyed. In other words, all mass entering a system will either leave that same system or will accumulate (build up) in the system. Writing this in formula form,

$$
\begin{matrix}
\text{Rate that} & & \text{Rate that} & & \text{Rate that mass} & \\
\text{mass enters} & = & \text{mass leaves} & + & \text{accumulates} & \quad (5.1) \\
\text{the system} & & \text{the system} & & \text{in the system} &
\end{matrix}
$$

We encounter this concept in our everyday lives, such as when we fill a bathtub. If we try to add water to a half-filled tub with the drain open, the following situations are possible:

situation #1: the water will enter faster than it leaves through the drain (the water level will rise - hence the water will accumulate)

situation #2: the water will enter slower than it leaves through the drain (the water level will fall - the accumulation will be negative)

situation #3: under the right circumstances, the water will enter at exactly the same rate as it leaves through the drain (the water level will not change - accumulation is zero and the system is at steady state)

Unlike total *mass*, total *moles* are not always conserved, because we sometimes have chemical reactions taking place which change the number of total moles (e.g. A + B —> C). Hence, an equation similar to Equation 5.1 cannot be written for total moles.

If the process with which we are dealing is also <u>steady state</u>, then nothing changes with time. For such a process, there would be no accumulation of mass in the system, because an accumulation of mass would be a change with time. For such a steady-state process, Equation 5.1 reduces to Equation 5.2.

$$
\text{Rate of mass entering system} = \text{Rate of mass leaving system} \quad (5.2)
$$

Let's examine how Equation 5.2 applies to a typical "system" (e.g. part of a process, a holding tank, etc.) such as the one depicted in Figure 5.1 in which several flowing streams bring material into the system and several streams take material out of the system.

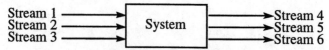

Figure 5.1 Schematic of a general "system" with input and output streams

For the system in Figure 5.1, the concept expressed in Equation 5.2 would be written

$$\dot{m}_1 + \dot{m}_2 + \dot{m}_3 = \dot{m}_4 + \dot{m}_5 + \dot{m}_6$$

where the units of each term are *mass/time*. A more general statement of this same principle is given in Equation 5.3, which is the formal statement of the <u>Steady-State Total Mass Balance</u>:

$$\sum_{\substack{input \\ streams}} \dot{m}_{in} = \sum_{\substack{output \\ streams}} \dot{m}_{out} \qquad\qquad (5.3)$$

where the subscript "*in*" in the left-hand summation indicates that the summation will include a mass flow rate for each of the input streams (e.g. the subscript might be replaced with a stream number as in Figure 5.1), and the subscript "*out*" similarly represents each of the output streams. While this idea is straightforward and may seem too simple to write in such formal terms, it is important for you to practice thinking in terms of Equation 5.3. The following example illustrates the approach:

Example 5.1

Three different streams deposit contaminated oil in a waste oil tank, and the mass flow rates of those streams are given below.

input stream	mass flow rate (lb_m/hr)
1	196.7
2	243.9
3	119.3

The waste oil tank sits on a scale, and an outlet pipe is placed in the tank so that the waste oil can be steadily withdrawn in an attempt to keep the scale reading from changing. At what mass flow rate must the oil be withdrawn to maintain a constant scale reading?

<u>Solution</u>: The diagram for this problem is as follows:

$$\dot{m}_1 = 196.7 \; lb_m/hr$$
$$\dot{m}_2 = 243.9 \; lb_m/hr$$
$$\dot{m}_3 = 119.3 \; lb_m/hr$$

waste oil → $\dot{m}_{out} \; lb_m/hr$

scale

Since the flow rates into and out of the tank will be constant with time and the mass in the tank will also be constant, we are dealing with a steady-state process. Therefore, Equation 5.3 becomes

$$\dot{m}_{out} = \dot{m}_1 + \dot{m}_2 + \dot{m}_3 = 196.7 + 243.9 + 119.3 = 559.9 \; lb_m/hr$$

The calculation in Example 5.1 may seem trivial, but the value of this formal approach will become more evident as we begin to deal with increasingly-complex ways to express the mass balance. For example, in Chapter 4, we learned that mass flow rate could also be expressed in terms of density and volumetric flow rate, as shown below (see also Equation 4.4).

$$\dot{m} = \rho\dot{V} \tag{5.4a}$$

For a stream of pure species A, $\dot{m} = \rho\dot{V} = MW_A\dot{n}$ (5.4b)

We repeat the encouragement given in Chapter 4 <u>that you fully understand and can reconstruct these alternate forms of mass flow rate</u>. We can substitute the equivalent forms into the steady-state mass balance to produce many variations, as follows (with Equation 5.3 repeated for comparison purposes):

$$\sum_{\substack{input \\ streams}} \dot{m}_{in} = \sum_{\substack{output \\ streams}} \dot{m}_{out}$$

$$\sum_{\substack{input \\ streams}} (\rho\dot{V})_{in} = \sum_{\substack{output \\ streams}} (\rho\dot{V})_{out}$$

$$\sum_{\substack{input \\ streams}} \dot{m}_{in} = \sum_{\substack{output \\ streams}} (\rho\dot{V})_{out}$$

$$\sum_{\substack{input \\ streams}} (\rho\dot{V})_{in} = \sum_{\substack{output \\ streams}} \dot{m}_{out}$$

where other variations include expressing the mass flow rate of one input stream using one form (e.g. $\dot{m}$) and another input stream using the other form (e.g. $\rho\dot{V}$). In each equation, each term (or product of terms) in each summation sign still has the units of *mass/time*.

To illustrate, let's re-visit Example 5.1, but with the information given in a different form:

Example 5.2

The three streams of Example 5.1 have the flow rates and densities given below.

input stream	volumetric flow rate (gal/hr)	density (lb_m/ft^3)
1	27.4	53.7
2	35.7	51.1
3	19.4	46.0

If the oil leaving the tank has a density of 50.8 lb_m/ft^3, at what volumetric flow rate must the oil be withdrawn to maintain a constant scale reading?

<u>Solution</u>: The diagram for this problem is as follows:

$$\frac{(27.4)(53.7) + (35.7)(51.1) + (19.4)(46.0)}{50.8 \ lb_m/ft^3} = \dot{V}$$

$\dot{V}_1 = 27.4\ gal/hr,\ \rho_1 = 53.7\ lb_m/ft^3$

$\dot{V}_2 = 35.7\ gal/hr,\ \rho_2 = 51.1\ lb_m/ft^3$

$\dot{V}_3 = 19.4\ gal/hr,\ \rho_3 = 46.0\ lb_m/ft^3$

waste oil scale $\quad \dot{V}_{out} = ?\ \rho_{out} = 50.8\ lb_m/ft^3$

Again, the mass balance for this system is

$$\dot{m}_1 + \dot{m}_2 + \dot{m}_3 = \dot{m}_{out}$$

but we can substitute forms for the mass flow rate which are more convenient (because of the forms of the data which are given). In this case, a more convenient form is

$$\rho_1 \dot{V}_1 + \rho_2 \dot{V}_2 + \rho_3 \dot{V}_3 = \rho_{out} \dot{V}_{out}$$

Rearranging,

$$\dot{V}_{out} = \frac{\rho_1 \dot{V}_1 + \rho_2 \dot{V}_2 + \rho_3 \dot{V}_3}{\rho_{out}}$$

$$= \frac{\left(53.7 \frac{lb_m}{ft^3}\right)\left(27.4 \frac{gal}{hr}\right) + \left(51.1 \frac{lb_m}{ft^3}\right)\left(35.7 \frac{gal}{hr}\right) + \left(46.0 \frac{lb_m}{ft^3}\right)\left(19.4 \frac{gal}{hr}\right)}{50.8 \frac{lb_m}{ft^3}} = 82.4 \frac{gal}{hr}$$

You should notice that a final algebraic solution was found before numbers were inserted into the solution and a calculation was made. This procedure (of finding the algebraic solution first) is recommended because it will help you

1. organize the problem you are trying to solve,
2. discover errors in your solution (because you will be able to see relationships between the algebraic symbols), and
3. avoid unnecessary calculations.

The conservation of total mass is a general principle which can be assumed to be valid in virtually all circumstances. Because of its general applicability, it is a universal basis for determining some unknown input or output flow rate or density from the other flow rates and densities. Even when other balances must be constructed to determine additional information (you learned in your mathematics courses that the number of equations must equal the number of unknowns), the total mass balance is often required along with the other balance(s) to produce the complete solution.

These first two examples are relatively simple, and the strategy for applying the general material balance equations is somewhat intuitive. However, material-balance problems can get sufficiently complex that our intuition may not be enough to get us through the solution. Here, it becomes helpful to have a stepwise strategy for solving such problems. Such a strategy is outlined in Table 5.1; you should discipline yourself to develop the habit of following this approach even for simple problems, so that you will instinctively do so for more complex problems where the strategy will become particularly helpful.

Table 5.1 Steps for Analyzing Material Balance Problems

1. Draw a diagram if one is not already available.

2. Write all *known* quantities (flow rates, densities, etc.) in the appropriate locations on the diagram. If symbols are used to designate known quantities, include those symbols on the diagram.

3. Identify and assign symbols to all *unknown* quantities and write them in the appropriate locations on the diagram.

4. If no flow rates are known, assume a convenient value for one of the flow rates as a *basis of calculation* (e.g. 100 *lbmol/s*, 100 *kg/hr*, etc.).

5. Determine the appropriate set of equations needed to solve for the unknown quantities. In order for the problem to be solvable, the number of equations must equal the number of unknowns. The steps below can be used to obtain the desired set of equations.

   a. Construct the material balance equation(s):
       1) Start with the general equation (so you don't forget something)
       2) Discard terms which equal zero in your specific problem
       3) Replace remaining terms with more convenient forms (because of given information or selected symbols)

   b. Construct equations to express other known relationships between variables (remember, the total number of equations must equal the number of unknowns).

6. Solve algebraically for the desired parameters and then determine their values.

The steps in Table 5.1 are illustrated in the following example:

Example 5.3

Your company uses a process to concentrate orange juice by freeze drying. The input to the process is orange juice which has a density of 1.01 g/cm^3. Two streams are output from the process. The first output stream is the orange juice concentrate. The second is an ice slurry which has a density of 0.93 g/cm^3. The orange juice is concentrated to the point where the volume of the concentrate is 1/4 that of the incoming juice. Also, from experience it is known that we want to remove ice slurry at the following volumetric rate (where J = juice and I = ice):

$$\dot{V}_I = 0.7 \frac{\rho_J \dot{V}_J}{\rho_I}$$

How much does 1.0 L of concentrate weigh?

Solution: We recognize that we can determine the weight of 1.0 L of concentrate by first finding the concentrate density.

Steps 1-3: Now we can draw the diagram, where we have one input stream (juice) and two output streams (ice slurry and concentrate). The diagram is shown below with the pertinent values and symbols included.

Juice

$\rho_J = 1.01 \ g/cm^3$

$\dot{V}_J$ (unspecified)

$100 \ cm^3/s$

Ice

$\dot{V}_I = ?$ $\rho_I = 0.93 \ g/cm^3$

Concentrate

$\dot{V}_c = ?$ $\rho_c = ?$

$\rho_J \dot{V}_J = \dot{V}_I \rho_I + \dot{V}_c \rho_c$

Step 4: No flow rates are given, so we must choose a basis of calculation. In this case, it is convenient to choose $\dot{V}_J = 100 \ cm^3/s$.

Step 5a: We now write a total mass balance for the system, which is:

$$\dot{m}_J = \dot{m}_I + \dot{m}_c$$

Substituting in forms of the mass flow rates which are more convenient (because of the given information)

$$\rho_J \dot{V}_J = \rho_I \dot{V}_I + \rho_c \dot{V}_c$$

Step 5b: The unknowns in this equation are: $\dot{V}_I$, $\dot{V}_c$, and ρ_c. Thus, we have a single equation and 3 unknowns. Additional relationships are then required. From the problem statement:

$$\dot{V}_I = 0.70 \frac{\rho_J \dot{V}_J}{\rho_I} \quad \text{and} \quad \dot{V}_c = 0.25 \dot{V}_J$$

These two relationships can be used to eliminate the two volumetric flow rates as unknowns, leaving a single equation (the mass balance) and a single unknown (ρ_c).

Step 6: We now solve our mass balance algebraically to obtain an expression for the desired density:

$$\rho_c = \frac{\rho_J \dot{V}_J - \rho_I \dot{V}_I}{\dot{V}_c}$$

Substituting in the above relationships:

$$\rho_c = \frac{\rho_J \dot{V}_J - \rho_I \left(0.70 \frac{\rho_J \dot{V}_J}{\rho_I} \right)}{0.25 \dot{V}_J}$$

$$= 4(\rho_J - 0.70 \rho_J) = 4(0.30 \rho_J) = 1.21 \rho_J = 1.21 \ g/cm^3$$

Now that we know the density of the concentrate, we need to determine the weight of one liter of concentrate. To do this, we first convert the density to units of mass (say *kg*) per *liter*.

$$\rho_c = 1.21 \frac{g}{cm^3} \left(\frac{1 \ kg}{1000 \ g} \right) \left(\frac{1000 \ cm^3}{1 \ L} \right) = 1.21 \ kg/L$$

We then calculate the mass by multiplying by 1 liter of volume:

$$\text{mass} = m = \rho V = (1.21 \ kg/L)(1 \ L) = 1.21 \ kg$$

Finally, we calculate the weight by multiplying by the acceleration of gravity.

$$\text{weight} = mg = (1.21 \ kg)(9.81 \ m/s^2) \left(\frac{1 \ N}{1 \ kg \, m/s^2} \right) = 11.8 \ N$$

Section 5.2 Material Balances for Multiple Species

A balance can also be constructed for a particular chemical compound. However, in contrast to total mass, which is neither created nor destroyed, a particular chemical compound, let's call it compound "A," can be "created" (formed) or "destroyed" (consumed by transforming it to another compound) by chemical reaction which occurs in the system. In that case, the formation of "A" is an additional input of "A" to the system, and the consumption of "A" is an additional output. Thus Equation 5.1 becomes

$$
\begin{array}{l}
\text{Rate that} \\
\text{A enters} \\
\text{the system}
\end{array}
+
\begin{array}{l}
\text{Rate that} \\
\text{A is formed} \\
\text{in the system}
\end{array}
=
\begin{array}{l}
\text{Rate that} \\
\text{A leaves} \\
\text{the system}
\end{array}
+
\begin{array}{l}
\text{Rate that} \\
\text{A is consumed} \\
\text{in the system}
\end{array}
+
\begin{array}{l}
\text{Rate that} \\
\text{A accumulates} \\
\text{in the system}
\end{array}
\quad (5.5)
$$

Again, we will restrict our focus to situations which are <u>steady state</u>, so nothing changes with time, and there would be no accumulation of mass of species A in the system. For such a steady-state process, Equation 5.5 would reduce to Equation 5.6.

$$
\begin{array}{l}
\text{Rate that} \\
\text{A enters} \\
\text{the system}
\end{array}
+
\begin{array}{l}
\text{Rate that} \\
\text{A is formed} \\
\text{in the system}
\end{array}
=
\begin{array}{l}
\text{Rate that} \\
\text{A leaves} \\
\text{the system}
\end{array}
+
\begin{array}{l}
\text{Rate that} \\
\text{A is consumed} \\
\text{in the system}
\end{array}
\quad (5.6)
$$

To formulate Equation 5.6 in terms that we can use, we introduce the following definitions:

$R_{formation,A}$ = rate that species A is formed, in units of <u>mass/time</u>

$R_{consumption,A}$ = rate that species A is consumed, in units of <u>mass/time</u>

$r_{formation,A}$ = rate that species A is formed, in units of <u>moles/time</u>

$r_{consumption,A}$ = rate that species A is consumed, in units of <u>moles/time</u>

From these definitions, it should be clear that

$$R_{formation,A} = (MW)_A r_{formation,A} \quad (5.7a)$$

$$R_{consumption,A} = (MW)_A r_{consumption,A} \quad (5.7b)$$

where $(MW)_A$ is the molecular weight of species A. Expressing Equation 5.6 with symbols as before, the <u>Steady-State Mass Balance for Species A</u> is

$$
\boxed{
\sum_{\substack{input \\ streams}} \dot{m}_{A,in} + R_{formation,A} = \sum_{\substack{output \\ streams}} \dot{m}_{A,out} + R_{consumption,A}
}
\quad (5.8)
$$

where all terms in the equation are in units of mass/time. The following example illustrates the use of Equation 5.8.

Example 5.4

Natural gas, which is principally methane, undergoes combustion by mixing it with air in a small burner. The following steady input and output streams apply to the burner:

Stream	Mass Flow Rate (g/s)	
	Methane	Other species
Natural gas input	4.61	0
Air input	0	not given
Output (Flue gas)	.09	not given

At what rate (in *g/s*) is methane being burned? (We are ignoring the other burnable materials in the natural gas.)

<u>Solution</u>: Because the flow rates and concentrations are all steady, we can assume that this is a steady-state process. We are looking for the rate at which methane is being burned or consumed. This is exactly what the term $R_{consumption,A}$ represents in Equation 5.8 (in this case $R_{consumption,methane}$). The process diagram would be as follows:

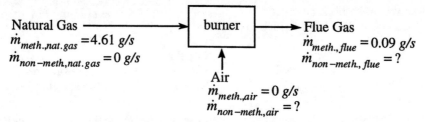

The methane mass balance is

$$\dot{m}_{meth.,nat.gas} + \dot{m}_{meth.,air} + R_{form.,meth.} = \dot{m}_{meth.,flue} + R_{cons.,meth.}$$

But no methane is being formed in the burner, so the rate of formation of methane equals zero, and the flow rate of methane in the air stream also equals zero. Equation 5.8 then reduces to

$$\dot{m}_{meth.,nat.gas} = \dot{m}_{meth.,flue} + R_{cons.,meth.}$$

Rearranging,

$$R_{cons.,meth.} = \dot{m}_{meth.,nat.gas} - \dot{m}_{meth.,flue}$$

$$= \ 4.61 \ g/s - 0.09 \ g/s = \ 4.52 \ g/s$$

In Equation 4.5 in Chapter 4, we saw that the mass flow rate for species A may be expressed in a number of ways, as repeated here in Equation 5.9.

$$\dot{m}_A = x_A \dot{m} = (MW)_A \dot{n}_A = (MW)_A y_A \dot{n} = (MW)_A c_A \dot{V} \qquad (5.9)$$

Again, we repeat that <u>it is important that you fully understand and can reconstruct these equivalent forms of mass flow rate</u>. For example, suppose that we are constructing a mass balance (Equation 5.8) on species A in a system. Suppose, further, that the mass flow rate of a stream ($\dot{m}$) is known, along with the mass fraction of species A in that stream (x_A). Under such circumstances, it would be convenient to express the mass flow rate of A in that stream ($\dot{m}_A$) as $x_A \dot{m}$. You should also remember that the sum of all mass fractions in a stream ($x_A + x_B + x_C + ...$) must equal 1.0, and the sum of all mole fractions ($y_A + y_B + y_C + ...$) likewise must equal 1.0.

Obviously, the mass balance for species A can take many forms, because the equivalent forms indicated in Equation 5.9 can be substituted into any of the terms in the balance as shown below, where the general form of the balance (Equation 5.8) is repeated here for comparison.

$$\underset{\substack{input \\ streams}}{\sum} \dot{m}_{A,in} + R_{formation,A} = \underset{\substack{output \\ streams}}{\sum} \dot{m}_{A,out} + R_{consumption,A}$$

$$\underset{\substack{input \\ streams}}{\sum} x_A \dot{m}_{in} + R_{formation,A} = \underset{\substack{output \\ streams}}{\sum} (MW)_A \dot{n}_{A,out} + R_{consumption,A}$$

$$\underset{\substack{input \\ streams}}{\sum} (MW)_A \dot{n}_{A,in} + R_{formation,A} = \underset{\substack{output \\ streams}}{\sum} (MW)_A (c_A \dot{V})_{out} + R_{consumption,A}$$

$$\underset{\substack{input \\ streams}}{\sum} (MW)_A (y_A \dot{n})_{in} + R_{formation,A} = \underset{\substack{output \\ streams}}{\sum} \dot{m}_{A,out} + R_{consumption,A}$$

$$\underset{\substack{input \\ streams}}{\sum} (MW)_A (c_A \dot{V})_{in} + R_{formation,A} = \underset{\substack{output \\ streams}}{\sum} x_A \dot{m}_{out} + R_{consumption,A}$$

etc.

etc.

You should note that <u>this equation is written for only one species at a time</u>! Thus, all of the molecular weight terms in the equation are for the same species and, therefore, are the same. Furthermore, all of the concentration, flow rate, and reaction terms are for species A only. One can write this balance for each species present, but again, each balance equation is written for only one species.

Let's try the previous example again, but with the data given in a different form.

Example 5.5

Natural gas, which is principally methane, undergoes combustion by mixing it with air in a small burner. The following steady input and output streams apply to the burner:

Stream	Flow rate (L/s)	Methane Conc. (gmol/L)
Natural gas input	7.2	0.04
Air input	unknown	0
Output (Flue gas)	58.4	0.0001

At what rate (in *g/s*) is methane being burned? (We are ignoring the other burnable materials in the natural gas.)

Solution: Because the flow rates and concentrations are all steady, we can assume that this is a steady-state process. The process diagram would be as follows:

Natural Gas ⟶ burner ⟶ Flue Gas
$\dot{V}_{nat.gas} = 7.2 \; L/s$ $\dot{V}_{flue} = 58.4 \; L/s$
$c_{meth.,nat.gas} = 0.04 \; gmol/L$ $c_{meth.,flue} = 0.0001 \; gmol/L$

Air
$\dot{V}_{air} = ?$
$c_{meth.,air} = 0 \; gmol/L$

Furthermore, no methane is being formed in the burner, so the rate of formation is zero. The mass balance on methane is

$$\dot{m}_{meth.,nat.gas} + \dot{m}_{meth.,air} + R_{form.,meth.} = \dot{m}_{meth.,flue} + R_{cons.,meth.}$$

But the rate of formation of methane equals zero (methane is not being formed), and the flow rate of methane in the air also equals zero, so the balance becomes

$$\dot{m}_{meth.,nat.gas} = \dot{m}_{meth.,flue} + R_{cons.,meth.}$$

Finally, it is more convenient to express the mass flow rate in terms of concentrations and volumetric flow rates, so we write

$$MW_{meth.}(c_{meth.,nat.gas}\dot{V}_{nat.gas}) = MW_{meth.}(c_{meth.,flue}\dot{V}_{flue}) + R_{cons.,meth.}$$

Rearranging,
$$R_{cons.,meth.} = MW_{meth.}(c_{meth.,nat.gas}\dot{V}_{nat.gas}) - MW_{meth.}(c_{meth.,flue}\dot{V}_{flue})$$

$$= (16 \; g/gmol)(0.04 \; gmol/L)(7.2 \; L/s) - (16 \; g/gmol)(0.0001 \; gmol/L)(58.4 \; L/s)$$

$$= 4.61 \; g/s - 0.09 \; g/s = 4.52 \; g/s$$

In problems involving a chemical reaction and where the stoichiometry of the reaction is known, it is usually more convenient to use _mole balances_ to solve the problem. To obtain those balances, we substitute Equations 5.7a and 5.7b into Equation 5.8 and express the mass flow rate as $(MW)_A \dot{n}_A$. The result is that the molecular weight of species A appears in all the terms of the material balance and can be divided out. This results in a <u>Mole Balance on Species A</u> (every term has units of moles/time):

$$\sum_{\substack{input \\ streams}} \dot{n}_{A,in} + r_{formation,A} = \sum_{\substack{output \\ streams}} \dot{n}_{A,out} + r_{consumption,A} \qquad (5.10)$$

Again, recognizing the equivalent forms for the molar flow rate of species A suggested by some of the terms in Equation 5.9, we can substitute those forms as is convenient (because of the forms of the given data or desired quantities), such as

$$\sum_{\substack{input \\ streams}} (y_A \dot{n})_{in} + r_{formation,A} = \sum_{\substack{output \\ streams}} \dot{n}_{A,out} + r_{consumption,A}$$

$$\sum_{\substack{input \\ streams}} (c_A \dot{V})_{in} + r_{formation,A} = \sum_{\substack{output \\ streams}} (y_A \dot{n})_{out} + r_{consumption,A}$$

These relations are particularly useful in cases where the stoichiometry of the chemical reaction is known, because the stoichiometry helps us relate the _molar_ rates of formation and/or consumption of the participating species. This will be illustrated later in this section.

When solving problems using material balance equations, it is frequently necessary to use more than one balance to solve a given problem. For example, a balance on total mass AND a balance on one of the species may both be needed to arrive at a unique solution. Or balances on two separate species (e.g. species A and species B) might be needed. In fact, one can write as many balance equations as there are species (one balance per species), or one can write balances on all species except one plus a balance on total mass. Of course, one can write equations representing additional information as well (e.g. given flow rates, given conversions, etc.). The strategy is to keep writing equations until the total number of equations equals the total number of unknowns.

You should appreciate that problems involving material balances, especially with multiple species, can become very complex, and the systematic approach outlined in Table 5.1 is vital to their solution, as illustrated in Examples 5.4 and 5.5. Some additional guidelines are helpful for solving problems involving multiple species and are presented in Table 5.2.

Table 5.2 Guidelines for Solving Material Balance Problems Involving Multiple Species

•Determine if species information is required, or if an overall mass balance will suffice. In general, these problems require species information.

•If information on a particular species is required, write the balance for that species first. It may be that a single-species equation will provide enough information to solve the problem.

•To simplify the problem, _if a chemical reaction with known stoichiometry is involved,_ use _species mole balances_ rather than mass balances.

•Do not attempt to balance the total number of moles for reacting systems if the reaction changes the number of moles.

•A total mass balance is frequently useful to determine a missing flow rate for systems where the densities of the input and output streams are approximately constant. The constant-density assumption is applicable to liquid systems which contain a small amount (small concentration) of a reactant or pollutant or dissolved substance such as a salt.

•Words like _consumed, formed, converted, reacted, produced, generated, absorbed, destroyed,_ etc. in the problem statement indicate that consumption or formation terms are required in the material balance. Systems which include chemical reactions always require formation and/or consumption terms.

•If a single species balance does not provide sufficient information to solve the problem, write additional material balances up to the total number of species. If there are still more unknowns than equations, look for additional relationships among the unknowns, such as
   a. given flow rates or ratios (e.g. "the incoming propane has a flow rate of 31 _kg/s_" or "16 moles of oxygen enter the reactor for each mole of octane that enters")
   b. fractions (mass or mole) of all species in a stream must add up to 1.0
   c. stoichiometry: if the process includes a chemical reaction, such as

$$\nu_A A + \nu_B B \rightarrow \nu_C C + \nu_D D$$

   where ν_A is the stoichiometric coefficient for species A, etc., then the reaction equation provides relationships between the values of the formation and consumption terms (expressed in moles/time) for the reacting species; e.g.

$$\frac{r_{consumption,B}}{r_{consumption,A}} = \frac{v_B}{v_A} \qquad \frac{r_{formation,C}}{r_{consumption,A}} = \frac{v_C}{v_A} \qquad \frac{r_{formation,D}}{r_{consumption,A}} = \frac{v_D}{v_A}$$

To use these relationships, select one of the species in the reaction to be a reference species and write the stoichiometric relationships relative to that reference, as is shown above where species "A" is used as the reference species.

d. conversion: if it is known that a certain fraction (X) of reactant A is converted (or "consumed") in the process, one can write that the rate of consumption of A equals that fraction of the total incoming flow rate of A, i.e.

$$R_{consumption,A} = X \sum_{\substack{input \\ streams}} \dot{m}_{A,in} \quad \text{OR} \quad r_{consumption,A} = X \sum_{\substack{input \\ streams}} \dot{n}_{A,in}$$

Note: Often the conversion is given as a percentage and must be converted to a fraction.

•Carry units as you work the problem. Calculation mistakes are frequently discovered as you try to work out the units.

Let's illustrate with some examples:

Example 5.6

Benzene and toluene (two similar compounds) are partially separated using a distillation column. The feed (input) stream of 100 *kg/hr* contains benzene at a mass fraction of 0.40 with the balance being toluene. In the overhead output stream, the benzene flow rate is 36 *kg/hr*, and in the bottoms output stream, the toluene flow rate is 54 *kg/hr*. What is the toluene flow rate in the overhead output stream and the benzene flow rate in the bottoms output stream?

<u>Solution:</u> The diagram with the pertinent information is as follows:

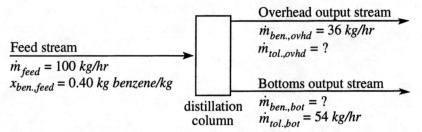

Because there are no chemical reactions, the formation and consumption terms in the material balance equations equal zero.

Benzene balance: $\dot{m}_{ben.,feed} = \dot{m}_{ben.,ovhd} + \dot{m}_{ben.,bot.}$

  or, more conveniently: $x_{ben.,feed}\dot{m}_{feed} = \dot{m}_{ben.,ovhd} + \dot{m}_{ben.,bot}$ (a)

Toluene balance: $\dot{m}_{tol.,feed} = \dot{m}_{tol.,ovhd} + \dot{m}_{tol.,bot}$

  or, more conveniently: $x_{tol.,feed}\dot{m}_{feed} = \dot{m}_{tol.,ovhd} + \dot{m}_{tol.,bot}$ (b)

Furthermore, we know that: $x_{ben.,feed} + x_{tol.,feed} = 1.0$ (c)

From Equation (c), $x_{tol.,feed} = 1.0 - x_{ben.,feed} = 1.0 - 0.4 = 0.6$

From Equation (a), $\dot{m}_{ben.,bot} = x_{ben.,feed}\dot{m}_{feed} - \dot{m}_{ben.,ovhd}$

$$= (0.4\ kg\ benzene/kg)(100\ kg/hr) - 36\ kg\ benzene/hr$$
$$= 4\ kg\ benzene/hr$$

From Equation (b), $\dot{m}_{tol.,ovhd} = x_{tol.,feed}\dot{m}_{feed} - \dot{m}_{tol.,bot}$

$$= (0.6\ kg\ benzene/kg)(100\ kg/hr) - 54\ kg\ benzene/hr$$
$$= 6\ kg\ toluene/hr$$

Example 5.7

As a design engineer, you are designing a process to convert some excess butene (C_4H_8) to ethylene (C_2H_4) using the reaction:

$$C_4H_8 \rightarrow 2C_2H_4$$

Your process will include a reactor and a separator, and two streams will exit the process: an ethylene-rich stream and a butene-rich stream. The following are requirements for the process:

Feed (input) stream: Flow rate: 865 lb_m/hr
                               Composition: Pure butene

Ethylene-rich output stream: Flow rate: 25.5 $lbmol/hr$
                               Composition: 92 $mole\%$ ethylene (the rest is butene)

Butene-rich output stream: Flow rate: 240 ft^3/hr
                               Composition: ethylene and butene

Of the incoming butene, 84% is converted to ethylene. What will be the concentrations of ethylene and butene (in units of $lbmol/ft^3$) in the butene-rich outlet stream?

Solution: The diagram and given information are as follows:

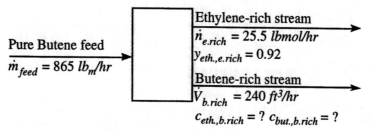

Given a chemical reaction with known stoichiometry, we will follow the recommendation to use mole balances.

Butene balance: $\dot{n}_{but.,feed} = \dot{n}_{but.,e.rich} + \dot{n}_{but.,b.rich} + r_{cons.,but.}$

or $\dfrac{\dot{m}_{but.,feed}}{(MW)_{but.}} = y_{but.,e.rich}\dot{n}_{e.rich} + c_{but.,b.rich}\dot{V}_{b.rich} + r_{cons.,but.}$ (a)

Ethylene balance: $r_{form.,eth.} = \dot{n}_{eth.,e.rich} + \dot{n}_{eth.,b.rich}$

or $r_{form.,eth.} = y_{eth.,e.rich}\dot{n}_{e.rich} + c_{eth.,b.rich}\dot{V}_{b.rich}$ (b)

Mole fractions: $y_{eth.,e.rich} + y_{but.,e.rich} = 1.0$ (c)

Stoichiometry: $\dfrac{r_{form.,eth.}}{r_{cons.,but.}} = \dfrac{2}{1} = 2$ (d)

Conversion: $r_{cons.,but.} = 0.84\,\dot{n}_{but.,feed} = \dfrac{0.84\,\dot{m}_{but.,feed}}{(MW)_{but.}}$ (e)

Finally, we need the molecular weights of butene and ethylene, which are easily determined from the chemical formulas:

$$(MW)_{eth.} = 12(2) + 4 = 28\ lb_m/lbmol$$
$$(MW)_{but.} = 12(4) + 8 = 56\ lb_m/lbmol$$

We now have 5 equations and 5 unknowns and can solve those equations, as follows:

From Equation (c): $y_{but.,e.rich} = 1.0 - y_{eth.,e.rich} = 1.0 - 0.92 = 0.08$

From Equation (e): $r_{cons.,but.} = \dfrac{0.84(865\,lb_m/hr)}{56\,lb_m/lbmol} = 13.0\,\dfrac{lbmol}{hr}$

From Equation (d): $r_{form.,eth.} = 2(r_{cons.but.}) = 26.0\ lbmol/hr$

From Equation (b): $c_{eth.,b.rich} = \dfrac{r_{form.,eth.} - y_{eth.,e.rich}\dot{n}_{e.rich}}{\dot{V}_{b.rich}}$

$$= \dfrac{26.0\,lbmol/hr - 0.92(25.5\,lbmol/hr)}{240\,ft^3/hr} = 0.0106\,lbmol/ft^3$$

From Equation (a): $c_{but.,b.rich} = \dfrac{\dfrac{\dot{m}_{but.,feed}}{(MW)_{but.}} - y_{but.,e.rich}\dot{n}_{e.rich} - r_{cons.,but.}}{\dot{V}_{b.rich}}$

$$= \dfrac{\dfrac{865\,lb_m/hr}{56\,lb_m/lbmol} - 0.08(25.5\,lbmol/hr) - 13.0\,lbmol/hr}{240\,ft^3/hr} = 0.0017\,lbmol/ft^3$$

We can now apply this approach to our problem of using NaOH to neutralize the HCl from our company process. The simplified diagram for this process was presented in Figure 3.2 as follows:

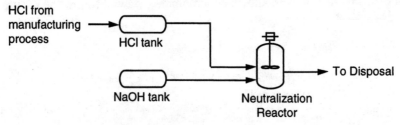

Figure 5.2 Process for neutralizing an HCl solution by the addition of a NaOH solution

As presented originally in Equation 3.1, the reaction of HCl and NaOH is

$$HCl + NaOH \rightarrow H_2O + NaCl \tag{5.11}$$

For this system, HCl and NaOH are consumed and not formed (so the $r_{formation}$ terms are zero). If we further require that the HCl and NaOH are totally consumed (their concentrations are zero in the stream leaving the reactor), the mole balances (with appropriate equivalent forms) are:

$$\text{HCl:} \quad \left(c_{HCl} \dot{V}_{HCl\ solution} \right)_{in} = r_{consumption,HCl} \tag{5.12}$$

$$\text{NaOH:} \quad \left(c_{NaOH} \dot{V}_{NaOH\ solution} \right)_{in} = r_{consumption,NaOH} \tag{5.13}$$

From the stoichiometry of the reaction, one mole of NaOH reacts with one mole of HCl, so:

$$\frac{r_{consumption,HCl}}{r_{consumption,NaOH}} = \frac{1}{1} = 1$$

or

$$r_{consumption,HCl} = r_{consumption,NaOH} \tag{5.14}$$

Solving Equations 5.12, 5.13 and 5.14 together gives, finally,

$$c_{NaOH} \dot{V}_{NaOH\ solution} = c_{HCl} \dot{V}_{HCl\ solution} \tag{5.15}$$

To compute the needed flow rate of NaOH solution,

$$\dot{V}_{NaOH\ solution} = \frac{c_{HCl} \dot{V}_{HCl\ solution}}{c_{NaOH}} \tag{5.16}$$

Equation 5.16 allows us to compute the necessary flow rate of NaOH solution for any combination of HCl and NaOH concentrations and flow rate of HCl solution. This will be useful in predicting the effects of changes of any of these variables.

You will see that we have accomplished at least two objectives. The principles of constructing material balances were outlined and applied to a number of kinds of problems. Those principles form the basis of many kinds of calculations in chemical engineering, including applications and configurations not mentioned in this introductory treatment. The second accomplishment was that the principles just mentioned were applied to our problem of neutralizing the HCl from our company process so that progress could be made in designing a new process to accomplish that neutralization. Those calculations, based on Equation 5.16, will be carried further in the next chapter.

READING QUESTIONS:

1. What assumptions are inherent in Equation 5.3? When do such assumptions <u>not</u> apply?

2. In what way(s) can a material balance on a particular chemical compound differ from a total material balance?

3. Generate an equation like Equation 5.9 to express the <u>molar</u> flow rate of a species, i.e. in the form $\dot{n}_A = ?? = ?? = ?? = ??$.

4. Explain the origin of Equation (e) in Example 5.7. What information in the problem statement was used in writing this relationship? What does $r_{consumption,butene}$ represent physically?

5. Suppose that you have a process which converts nitrogen and hydrogen into ammonia by the following chemical reaction:

$$N_2 + 3H_2 \rightarrow 2NH_3$$

a. For this process, is it correct to write a total mole balance such as $\sum \dot{n}_{in} = \sum \dot{n}_{out}$? Why or why not?

b. What is the maximum number of material balance equations that can be properly written for this process assuming that nitrogen, hydrogen and ammonia are the only chemical species present?

c. Eugene engineer claims that a total mass balance can be used along with three species balances for a total of four material balance equations. Do you agree? Why or why not?

6. Suppose that you (as a chemical engineer) have been asked to design a treatment facility for a hazardous waste stream. How might material balances be used to help you with your design?

HOMEWORK PROBLEMS:

1. Feed water is fed to a large steady-state boiler where most of it is converted to high-pressure steam, with a smaller amount of hot residual water discharged to waste. The water densities and flow rates and the steam density are as follows:

Stream	Density (kg/m^3)	Volumetric Flow rate (m^3/min)
Feed water	1000	28.0
Hot residual water	960	6.5
High-pressure steam	3.7	

What is the volumetric flow rate of the steam?

2. To make an industrial solvent, benzene (C_6H_6) is provided at a molar flow rate of 1140 *kgmol/hr.* Toluene (C_7H_8) is also added to the benzene at a rate of 213 *kgmol/hr* to enhance the solvent properties. Finally, it is necessary to add phenol (C_6H_6O) so that the final solvent production rate (mass flow rate) is 115,000 *kg/hr*. At what mass flow rate should the phenol be added?

3. In a candy company, separate streams of sugar, butter, corn syrup, vanilla extract, and milk enter a mixer-boiler and come out as fudge. The sugar (sucrose, $C_{12}H_{22}O_{11}$) is purchased from a sugar farmer who used to be a chemist and who packages it by the *lbmol*, and the process uses 1.75 *lbmol/hr*. Butter is fed to the process at a rate of 60 *lb_m/hr*, and corn syrup is fed at a rate of 3.5 *gal/hr*. Vanilla extract is fed at a rate corresponding to 1 *lb_m* of extract for every 30 *lb_m* of sugar. How many gallons of milk per hour must be fed to the process for a total fudge production of 830 *lb_m/hr*? (Assume that both the corn syrup and milk have densities equal to 62.4 *lb_m/ft^3*.)

4. A section of river receives heavy rainfall and is prone to overflow its banks. Suppose that the volumetric flow rate of water entering the section from upstream is "U." Further, the volumetric rate at which rainfall adds to that section of the river is "R." Finally, the volumetric flow rate leaving that section by flowing downstream is proportional to the height of the river in that section, or

$$D = k \cdot \text{Height}$$

a. Assuming that the height of the river in that section is steady, derive an algebraic expression for that height in terms of the given symbols.

b. If the height of the river bank is h_b, derive an algebraic expression (in terms of the given symbols) for the rate of rainfall that would just cause overflow.

5. Two streams of similar liquid mixtures enter the center of a distillation column. The action of the column separates the combined mixture into several components of different volatilities (and different densities), which leave the column at the top (product stream 1), middle (product stream 2), and bottom (product stream 3). Given the following data, what is the density of product stream 2?

stream	mass flow (kg/hr)	volumetric flow (m^3/hr)	density (kg/m^3)
feed 1	260,000		
feed 2		283	935
prod. 1		157	721
prod. 2		235	
prod. 3	208,000		

6. Because of environmental considerations, acetone must be removed from air used in your chemical plant before the air is released into the atmosphere. The acetone is removed by absorbing it into water and distilling the water to produce an acetone-rich stream and a water-rich stream.

Two streams enter the process as listed in the table below. The first is the dirty air stream, which has a total mass flow rate of 3500 kg/hr. The second is a pure water stream.

Three streams exit the process as shown in the table, a pure air stream, an acetone-rich liquid, and a water-rich liquid. The flow rate of the water-rich outlet stream has been measured to be 652 kg/hr.

What are the mass flow rates of the other streams?

	Dirty air in	Pure water in	Pure air out	Acetone-rich liquid out	Water-rich liquid out
x_{air}	0.984	0	1.0	0	0
$x_{acetone}$	0.016	0	0	0.989	0.042
x_{water}	0	1.0	0	0.011	0.958

7. One method of producing drinkable water is to filter sea water. A large facility continuously filters 2300 L of sea water per hour. The sea water contains sodium chloride (NaCl) at a concentration of 1.4 M. Two streams emerge from the facility: 1) a "brine" stream with a sodium chloride concentration of 5.6 M and flow rate of 560 L/hr and 2) a drinking water stream. The densities of the three streams can be considered to be the same. What is the sodium chloride concentration in the drinking water?

8. Human lungs transfer oxygen from air to blood in a process that can be considered to be steady state. Air enters the lungs containing a large amount of oxygen (a non-reacting species in that it is not formed or consumed in the lungs) and leaves the lungs with less oxygen. Meanwhile, the blood enters the lungs with less oxygen and leaves with more. From the oxygen concentrations and the blood flow rates listed below, what flow rate of air must pass through the lungs? The flow rate of air can be considered to be the same for the inspired (entering) air as for the expired (exiting) air.

Stream	O_2 Concentration (gmol/L)	Flow rate (L/min)
Entering air	0.00934	
Exiting air	0.00705	
Entering blood	0.00670	5.0
Exiting blood	0.00893	5.0

9. Gas in a tank contains a poison at a concentration c_T. It was just discovered that the gas is leaking into the surrounding room at a rate of $T \, cm^3/min$. Meanwhile, an air conditioner brings fresh air into the room at a rate of $A \, cm^3/min$, and the air in the room (which is well-mixed) leaves through an open window. The densities of all the gases are the same. Initially, the concentration of poison in the room (c_R) will rise, but it will eventually reach a steady value. In terms of the given symbols, what is that steady concentration?

10. A manufacturing plant located in a narrow canyon has applied to the Environmental Protection Agency (EPA) for a permit to continuously release 15 ft^3/min of exhaust gas, containing 0.06 $gmol/ft^3$ of a pollutant, into the air. The wind travels through the canyon at an average flow rate of 800 ft^3/min (not including the exhaust gas). Furthermore, the pollutant also reacts with the air and sunlight to be converted to a harmless product (the volume of air consumed in this reaction is negligible) so that 20% of the pollutant released is converted by the time it leaves the canyon.

   a. What will be the average concentration of the pollutant in the air coming out of the canyon?

   b. What wind flow rate entering the canyon would produce a pollutant concentration greater than 0.0025 $gmol/ft^3$ in the air coming out of the canyon?

   Hint: The densities of the exhaust gas and the air can be assumed to be the same.

11. For the combustion of methane presented in Examples 5.4 and 5.5, the chemical reaction is

$$CH_4 + 2O_2 \rightarrow CO_2 + 2H_2O$$

   Let's suppose that methane flows into a burner at 30 $gmol/s$, while oxygen flows into the same burner at 75 $gmol/s$. If all of the methane is burned, and a single output stream leaves the burner, what is the mole fraction of CO_2 in that output stream?

   Hint 1: Does the fact that all the methane is burned mean that all the oxygen is burned also?
   Hint 2: Find the molar flow rate of each component gas in the flue gas.

12. A stream of pure species A (50 $kgmol/hr$) and a stream of pure species B (5 $kgmol/hr$) enter a reactor where the A is converted to B according to the following reaction:

$$2A \rightarrow B$$

If the molar flow rate of A in the output stream from the reactor is 20 $kgmol/hr$, what is the molar flow rate of species B in that same output stream?

13. A chemical process is used to convert toluene ($MW=92$) and hydrogen to benzene ($MW=78$) and methane (CH_4) by the following reaction:

$$\text{Toluene} + H_2 \rightarrow \text{Benzene} + CH_4$$

Two streams enter the process. The first input stream is pure liquid toluene, which enters at a rate of 40 $gmol/s$. The second input stream is a mixture of H_2 (95 $mole\%$) and CH_4 (5 $mole\%$). The flow rate of H_2 in the second stream is equal to 200 $gmol/s$.

Two streams leave the process. The first output stream contains only liquid benzene (product) and toluene (unconverted reactant). The second output stream contains gaseous H_2 and CH_4.

If the conversion of toluene in the process is 75% (i.e. 25% is unreacted), what are the molar flow rate and mass fraction of benzene in the liquid output stream?

Hint: Like many situations in life, this problem may provide more information than you need to answer the given question.

14. Hydrogen is a very clean fuel, because it burns to form water with no pollutants according to the reaction

$$2H_2 + O_2 \rightarrow 2H_2O$$

Pure hydrogen gas is fed to a burner at a rate of 38 $gmol/min$. Also fed to the burner is a stream of air, which has an oxygen mole fraction of 0.21, with the balance being nitrogen. The flow rate of the air is 10% greater than would be needed to exactly react with all of the hydrogen (we call this "10% excess air"). If the output gas (flue gas) from the burner contains no hydrogen, what are the molar flow rates of all other constituents in the output gas?

15. A steady-state chemical process is used to convert nitrogen (N_2) and hydrogen (H_2) to ammonia (NH_3) by the following reaction:

$$N_2 + 3H_2 \rightarrow 2NH_3$$

Stream 1, containing 95 $mole\%$ nitrogen and 5 $mole\%$ hydrogen, enters the process at a rate of 400 $lbmol/hr$, and stream 2, containing pure hydrogen (density = 0.08 lb_m/ft^3), enters the process at a volumetric flow rate of 31,000 ft^3/hr. A single stream leaves the process. If all of the nitrogen is consumed in the reaction, what is the molar flow rate of hydrogen in the exiting stream?

Courtesy of Kraft Foods, Inc., North Lawrence, NY

CHAPTER 6

SPREADSHEETS
(CALCULATING THE COST OF THE BASE)

Section 6.1   The Calculation Scheme

Now that you have decided to neutralize the acid with sodium hydroxide, one of the next steps is to specify the flow rate and concentration of NaOH needed and to determine the cost associated with that NaOH. Let's suppose that you have identified the possible suppliers of sodium hydroxide and have determined that one of them (the XYZ Company) will provide it at the lowest cost and provides the product in various concentrations as shown in Table 6.1.

Table 6.1 Available NaOH Solutions from the XYZ Company

Preparation	Concentration	Preparation	Concentration
A	5.0 mM	K	30.0 mM
B	7.5	L	32.5
C	10.0	M	35.0
D	12.5	N	37.5
E	15.0	O	40.0
F	17.5	P	42.5
G	20.0	Q	45.0
H	22.5	R	47.5
I	25.0	S	50.0
J	27.5	T	52.5

In the table, the concentration of NaOH is given in units of milli-molar ($10^{-3}M$, abbreviated "mM"). Furthermore, the company tells us that the purchase price follows the price formula:

$$\$/L = 0.0058 \, (c_{NaOH})^{1.4} + 0.017 \qquad (6.1)$$

where c_{NaOH} is the concentration of the NaOH in units of mM. In those same units, the concentration of the acid (c_{HCl}) is 14.0 mM (see Chapter 3).

We know that pumping the NaOH solution will cost something for the electricity to operate the pump. Let's suppose we determine that, for higher (≥ 13 mM) NaOH concentrations, a chemically-inert tank and pump must be used in the delivery system to avoid chemical reaction between the NaOH and the metal, and that pump will be more expensive to operate. Finally, let's suppose that a relatively-simple pumping cost analysis tells us that the NaOH pumping cost is

$$\$0.173/L \qquad \text{for } c_{NaOH} < 13 \, mM$$
$$\$0.194/L \qquad \text{for } c_{NaOH} \geq 13 \, mM$$

where the discontinuous nature of this function is due to the need to switch to the chemically-inert tank and pump when c_{NaOH} is 13 mM or higher.

We would like to find the total cost for various possible NaOH concentrations so that we can find the concentration which produces the minimum total cost. We would do this as follows:

1. Determine the <u>needed NaOH flow rate</u> from a material balance. In Chapter 5, we applied the material balances to our problem of acid neutralization and derived the relation (Equation 5.16):

$$\dot{V}_{NaOH\ solution} = \frac{c_{HCl}\dot{V}_{HCl\ solution}}{c_{NaOH}} \qquad (6.2)$$

where the expressions c_{NaOH} and c_{HCl} refer to the molar concentrations of NaOH and HCl, respectively, and $\dot{V}_{NaOH\ solution}$ and $\dot{V}_{HCl\ solution}$ refer to the volumetric flow rates (volume per hour) of NaOH and HCl (from Chapter 3, the HCl flow rate is 11,600 *L/hr*).

2. Determine the <u>purchase cost</u> of the NaOH solution by determining the <u>purchase price</u> per liter from the price formula given above and then multiplying this price by the <u>volumetric flow rate</u> of NaOH just calculated.

3. Calculate the NaOH <u>pumping cost</u> by determining which pumping cost rate applies for the NaOH concentration being evaluated. This cost rate must also be multiplied times the required flow rate of NaOH.

4. Determine the <u>total cost</u> by adding the purchase cost of the NaOH to the pumping cost.

From your experience, you may think of several possible strategies for performing this series of calculations. Let's explore a few of them.

•**Hand-written calculation sheet**: One obvious strategy would be simply to use a calculator to perform the computations and to record the results on a piece of paper. Years ago, nearly all engineering calculations were done this way. Various NaOH concentrations would be organized in a column on the paper and then intermediate and final values would be calculated and recorded in separate columns (Figure 6.1). The advantage of this method is that one can easily see how the calculated values are affected by the independent variable (NaOH concentration, in this case). However, this work is tedious and must be redone each time a parameter changes.

•**Algebraic cost function**: The fact that we are looking for a minimum may have suggested to you that you use calculus to find that minimum. This would require that you derive an algebraic expression for the total cost as a function of the NaOH concentration. You would then take the derivative of that function, set the derivative equal to zero, and solve for the value of the independent variable. In cases where this method can be used, it produces an analytic solution which can be evaluated for any values of the independent variables. However, this method does not work as well for the problem at hand, because we are dealing with a function which does not change smoothly but is discontinuous (takes a sudden jump in value), due to the discontinuous nature of the pumping cost.

•**Computer program**: You may have had experience writing computer programs in a language like BASIC, PASCAL, or C and may now recognize that such a program could be written to perform the calculations described above. This would involve writing commands which cause the computer to input the needed information, make the necessary calculations, and output the results to a file, screen, or printer. The resulting program would allow you to vary the values of the input parameters or to extend the calculation to varying conditions. However, the process of writing the computer program can be somewhat involved and time-consuming. For example, one time-consuming aspect would be to have the program display of all of the intermediate results so that you could have a clear picture of the various contributions to the overall results.

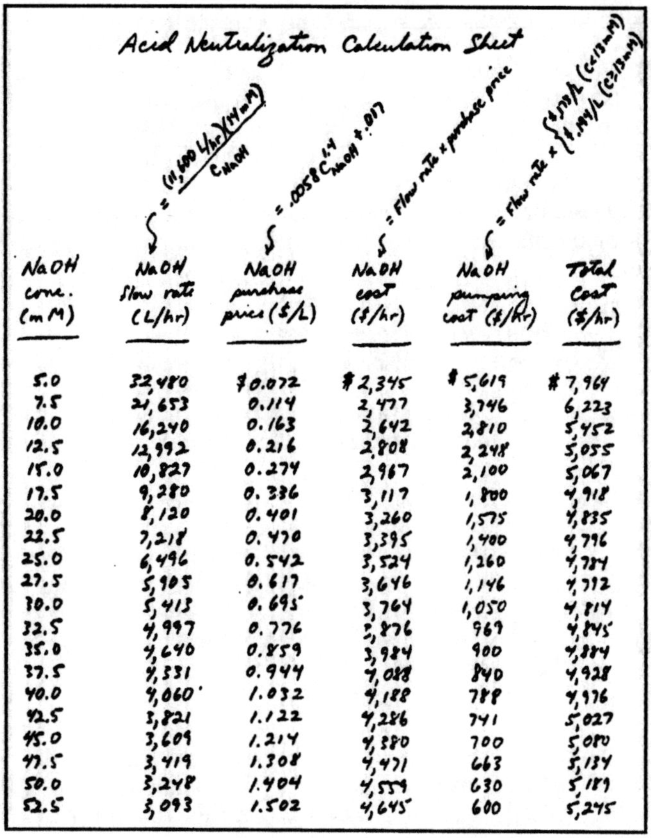

Acid Neutralization Calculation Sheet

NaOH conc. (mM)	NaOH flow rate (L/hr)	NaOH purchase price ($/L)	NaOH cost ($/hr)	NaOH pumping cost ($/hr)	Total Cost ($/hr)
	= (11,600 L/hr)(14 mM) / C NaOH	= .0058 C NaOH^1.4 + .017	= Flow rate x purchase price	= Flow rate x {$.573/L (C<13mM); $.194/L (C≥13mM)}	
5.0	32,480	$0.072	$2,345	$5,619	$7,964
7.5	21,653	0.114	2,477	3,746	6,223
10.0	16,240	0.163	2,642	2,810	5,452
12.5	12,992	0.216	2,808	2,248	5,055
15.0	10,827	0.274	2,967	2,100	5,067
17.5	9,280	0.336	3,117	1,800	4,918
20.0	8,120	0.401	3,260	1,575	4,835
22.5	7,218	0.470	3,395	1,400	4,796
25.0	6,496	0.542	3,524	1,260	4,784
27.5	5,905	0.617	3,646	1,146	4,792
30.0	5,413	0.695	3,764	1,050	4,814
32.5	4,997	0.776	3,876	969	4,845
35.0	4,640	0.859	3,984	900	4,884
37.5	4,331	0.944	4,088	840	4,928
40.0	4,060	1.032	4,188	788	4,976
42.5	3,821	1.122	4,286	741	5,027
45.0	3,609	1.214	4,380	700	5,080
47.5	3,419	1.308	4,471	663	5,134
50.0	3,248	1.404	4,559	630	5,189
52.5	3,093	1.502	4,645	600	5,245

Figure 6.1 Example of a hand-written calculation sheet which was used frequently in the past to record the results of engineering calculations

•**Spreadsheet**: Spreadsheets are computerized versions of the old hand-written calculation sheets. Furthermore, they perform the calculations as well as display them, so they can be constructed much more quickly and easily than the hand-written version. Intermediate and final values are easily determined and displayed using mathematical relationships constructed by the user for each row or column of numbers. Modern spreadsheet programs also include the capability to display the numbers graphically. Additionally, the spreadsheet can be set up so that values are instantly recalculated and replotted when one of the input values is changed. For all of these reasons, the spreadsheet is a popular tool for practicing engineers.

Section 6.2 Setting Up a Spreadsheet

In this section, we will talk about the characteristics of a spreadsheet and how to begin using one. The spreadsheet we will use as an example is Microsoft™ Excel, but other spreadsheets are also available. All spreadsheets are operated in a fashion which is similar to that of Excel, but the operating procedures and instruction sets will differ with each spreadsheet program.

The first step to using a spreadsheet like Excel will be to gain access to a computer which has the spreadsheet program stored on the computer's own storage device (usually a "hard disc") or which is connected via a network to another computer from which the spreadsheet can be accessed and run. You should be familiar with how to use a personal computer, whether of the type which uses Microsoft™ Windows or the Macintosh™ operating system. Of particular importance, you should know how to find a file and open it. In addition, you should know how to use the mouse, menus, and other features of the graphical operating system (called a "Graphical User Interface" or GUI). If needed, you are encouraged to go through the tutorial on Microsoft Windows or on the Macintosh to become familiar with how to use windows, scroll bars, menus, and icons.

Excel has online instructions which explain how to use the program. Those instructions are found under the "Help" pull-down menu which appears at the top of the screen when you open the Excel program. Some of the instructive information is designed for the beginner, and you are encouraged to access and read that material and to practice using the procedures described there. The discussion which follows will build upon the information in those instructions.

From the instructions under the Help menu, you will learn that each cell of the spreadsheet has a unique address, consisting of the column, designated by a letter, and the row, designated by a number (e.g. G9). You will also learn about "fill down" and "fill right" commands, which duplicate the information in one cell to other cells below and to the right of the original cell. If the information in the original cell also contains addresses of other cells, the "fill" feature changes those addresses; for the "fill down" feature, it increments the row number, and for the "fill right," it increments the column letter. This is important in applying the spreadsheet to the solution of our problem, as we will now demonstrate.

You will also learn that you can instruct Excel to perform mathematical operations. Some of the symbols and syntax for those instructions are listed in Table 6.2. For the first five operations in the Table, an "order of operation" is listed indicating that in a mathematical expression containing more than one of these operations, the raising to a power would be performed first, followed by multiplication and division, with addition and subtraction operations performed last. For example, if we wanted to compute a value from the equation

$$x = \frac{5^{3.7}}{\sqrt{117}} - 14.9$$

and wanted to place the results of that calculation (i.e. the value of x) into a particular cell in Excel, we would enter into that cell the formula

$$=5\text{^}3.7/\text{SQRT}(117)-14.9$$

This formula begins with an "equals" sign to tell Excel that what follows is an executable formula.

In the formula just given, Excel would perform the exponentiation before the division and would not place the square root in the exponent (see Note 2 in Table 6.2 describing the order of operations). It would also perform the division before the subtraction. Use parentheses to avoid ambiguous functions such as

$$\cos(1.93)^{2.6}$$

which could mean either of the following two functions:

Algebraic expression	Excel expression
$\cos[(1.93)^{2.6}]$	$\text{COS}(1.93\text{^}2.6)$
$[\cos(1.93)]^{2.6}$	$(\text{COS}(1.93))\text{^}2.6$

Table 6.2 Some Selected Intrinsic Functions for Microsoft EXCEL

		Order of operation
x^y	raises x to the y power	1
x*y	multiplies x and y	2
x/y	divides x by y	2
x+y	adds x and y	3
x-y	subtracts y from x	3

SIN(x)	sine of angle x (in radians)
COS(x)	cosine of angle x (in radians)
TAN(x)	tangent of angle x (in radians)
ASIN(x)	arc sine (angle is given in radians)
ACOS(x)	arc cosine (angle is given in radians)
ATAN(x)	arc tangent (angle is given in radians)

PI() returns the value of pi

ABS(x)	absolute value of x
EXP(x)	e^x
INT(x)	integer value of x (rounding down to the nearest integer)
LN(x)	natural logarithm of x
LOG10(x)	base-10 logarithm of x
SQRT(x)	square root of x

MAX(a:b)	maximum value in cells in the array a:b
MIN(a:b)	minimum value in cells in the array a:b
SUM(a:b)	summation of values in cells in the array a:b
AVERAGE(a:b)	average of values from cells in the array a:b
STDEV(a:b)	standard deviation of values in cells in the array a:b (assumes data represent a sample of a larger population)
VAR(a:b)	variance of values in cells in the array a:b (assumes data represent a sample of a larger population)

IF(expr1,expr2,expr3) places expression 2 or expression 3 in the cell, depending upon whether expression 1 is true or false
•expression 1 is a logical expression such as x>y
•expressions 2 and 3 may be text (in quotes), a number, or a formula to be executed

Notes:

1. The arguments x and y can be numbers or addresses of cells in which the numbers are found.

2. In a formula containing more than one math operation (addition, subtraction, multiplication, division, exponentiation), the computer will compute the exponentiations first, then the multiplications and divisions (in order from left to right), and then the additions and subtractions (in order from left to right). If an operation is to be performed on an argument which is the result of other operations, e.g. $[20.6 + 13 \cdot 4.2]^{1.7}$ or $\sin(16.8/1.2^3)$, the argument will be computed first.

In Excel, we can also write algebraic functions where the numeric values in those functions are the contents of cells. In the example just presented, we might construct the formula

$$=A7\text{^}F14/SQRT(B7)-H9$$

where A7, F14, B7 and H9 are cells containing the values to be used in the formula.

Setting up the Spreadsheet for the Acid-Neutralization Problem

The description which follows will be best understood by performing these steps on a computer while reading about them.

Let's begin constructing the spreadsheet for our problem by entering some headings. In cell C1, let's enter an overall title "NaOH Costs" and change it to Bold style using the Bold tool in the toolbar. Notice that the title is larger than the cell, but since nothing is entered into the adjacent cell to the right, the title is allowed to overflow into that adjacent cell. Let's also put the headings for the columns in cells A2 through F2. Finally, we'll enter the appropriate units below each heading (in cells A3 through F3). Using the hand-written calculation sheet (Fig. 6.1) as a guide, we'll choose the labels and units shown in Figure 6.2. You probably noticed that, unless told otherwise, Excel left-side justifies text (positions text along the left edge of the cell), so you may want to center these headings in the cells by selecting the cells and then using the "Alignment" item under the "Format" menu or by simply clicking on the center-justified tool in the toolbar.

To improve the appearance of the spreadsheet, you will want to adjust some column widths. You can do this by selecting each of those columns and using the "Column Width..." item under the "Format" menu or by pointing with the mouse at the edge of the header for the column (where the "letter" for the column is indicated) and "dragging" the edge of the column to make it wider or narrower. For example, you may want to set the widths of columns A through L as indicated in Table 6.3 (you'll see later why these were selected). The type of lettering (called the "Font") and size of that lettering also affect how well the type fits into the cells; you may want to choose a fairly compact combination of those features, such as a Times font, size 10 point, or Helvetica font, size 9 point.

	A	B	C	D	E	F
1			**NaOH Costs**			
2	Conc.	Flow	Price	Cost	Pmp Cost	Tot Cost
3	(mM)	(L/hr)	($/L)	($/hr)	($/hr)	($/hr)
4						
5						

Figure 6.2 The calculation headings

Table 6.3 Column Widths for the Worksheet

column	width	column	width
A	4	G	1
B	6	H	12
C	6	I	5
D	7	J	3
E	7	K	1
F	7	L	4

NaOH Concentrations (Column A):

Now let's enter the available concentrations of sodium hydroxide in column A. In cell A5, we'll enter the lowest concentration (5.0). We'll also set the number format by using the "Number" item under the "Format" menu to display one digit to the right of the decimal point by creating a format code "0.0" in the field next to the word "Code" and clicking the "OK" button. We can also set the format by using the decimal control tools on the tool bar.

We could type in each of the other NaOH concentrations, starting with cell A6, but there is a much easier way. The concentrations increase in increments of 2.5 *mM*, so we can use a formula. First, we select **cell A6** and type the formula

$$=A5+2.5$$

Don't forget to click on the "enter" button on the screen (the one with the check mark) or to press the Enter key on the keyboard. Now we select cells A6 through A24 and then click on "Fill Down" under the "Edit" menu (or use the autofill feature by dragging the "handle" from A6 through A24). Excel should have entered all of the concentrations. You will notice that a new formula was entered into each cell from A7 to A24; you can view each formula by clicking on each cell and reading the contents of the formula bar. Notice also that all the formulas are similar, except that the "Fill Down" function incremented the address in the formula by one row each time it moved to the next row.

Let's also display the given values of the HCl concentration and flow rate, the NaOH price formula, and the NaOH pumping cost in the spreadsheet. This will not only allow us to refer to them easily but will allow us to base our calculations on values which can easily be changed if necessary. Enter the information in columns H through L as shown in Figure 6.3.

	A	B	C	D	E	F	G	H	I	J	K	L
1			NaOH Costs									
2	Conc.	Flow	Price	Cost	Pmp Cost	Tot Cost		HCl flow (L/hr) =	11600			
3	(mM)	(L/hr)	($/L)	($/hr)	($/hr)	($/hr)		HCl conc (mM) =	14			
4												
5	5.0							NaOH cost ($/L)=				
6	7.5								0.0058	* C^	1.4 +	.017
7	10.0											
8	12.5							Pumping cost ($/L)=				
9	15.0							C<13mM	.173			
10	17.5							C>=13mM	.194			
11	20.0											

Figure 6.3 Spreadsheet with the NaOH concentrations, the adjusted columns, and the given data for HCl flow, HCl concentration, NaOH cost, and pumping costs

In cells H6 through L6, the information was entered in a way that suggests a formula to the observer. While Excel only recognizes those items as separate words and numbers in individual cells, this format makes it easy for you to keep track of the values and relationships and to retrieve them for your calculations.

NaOH Flow rate (Column B):

In column B, using Equation 6.2 from the previous section, we will calculate the NaOH flow rate required for each NaOH concentration we might purchase. Thus, we want to calculate

$$\dot{V}_{NaOH\ solution} = \frac{c_{HCl}\dot{V}_{HCl\ solution}}{c_{NaOH}} \qquad (6.2)$$

Click on **cell B5,** highlighting that cell. Everything we type on the keyboard is now tentatively entered into that cell. First we type an "equals" sign (=) indicating that we wish to enter a function or formula (i.e. an algebraic expression which calculates a value). The entire function that we type is

=I3*I2/A5

followed by the Enter key. In this formula, I3 represents the cell in which the concentration of HCl is found, I2 is where the HCl flow rate is found, and A5 is where the NaOH concentration is found. The value 32480 appears in the cell. We could have entered this formula even faster by clicking on each cell that we want entered into the formula, i.e. by typing "=" and then clicking on cell I3, then typing "*", then clicking on cell I2, then typing "/" and finally clicking on cell A5.

If we were to perform the "Fill Down" function for cells B5 through B24 (by either selecting those cells and using the Fill Down feature under the Edit menu or by pulling the "handle" on cell B5 down to B24), something would not work correctly (try it and see). That's because the Fill Down feature would increment every address in the formula found in cell B5, i.e. the formulas would be as follows in the cells:

cell	formula
B5	=I3*I2/A5
B6	=I4*I3/A6
B7	=I5*I4/A7
B8	=I6*I5/A8
etc.	etc.

Obviously, we don't want to increment the cell addresses I2 and I3 where the acid flow rate and concentration are found. To prevent this incorrect incrementing, we can "fix" those addresses in the formula in cell B5 by typing a dollar sign ($) in front of the number in those addresses, so the address reads

=I$3*I$2/A5

The dollar sign tells Excel not to increment the row number which follows the dollar sign. We could also put a dollar sign in front of the letter "I" to prevent the incrementing of the column letter for a Fill Right function also, but we won't be using a Fill Right function, so it won't be necessary.

Notice that we didn't place a dollar sign in front of the number in the reference to cell A5, because we want the row number of the NaOH concentration to be incremented. That is, we want the NaOH flow rate calculated in cell B6 to use the NaOH concentration in cell A6, etc. Now when we perform the Fill Down operation from cell B5 to B24, the correct numbers will be calculated. You will also want to set the format of those numbers by selecting them and using the format #,##0, meaning that we don't want any numbers to show to the right of the decimal place and we want commas to appear when the number equals 1,000 or larger.

NaOH Price (Column C):
Column C will contain the prices for each concentration of NaOH according to the price formula provided by the company, which is

$$\$/L = 0.0058\ C^{1.4} + 0.017$$

where C is in units of *mM*. To calculate the cost in column C, build the following formula in **cell C5**:

$$=H\$6*A5^J\$6+L\$6$$

where the references to H$6, J$6, and L$6 retrieve the values in the cost formula on the spreadsheet, and the dollar signs prevent those reference row numbers from being incremented when the Fill Down function is used. Now Fill Down the formula from C5 to C24 and change the Format of the numbers to one which shows a dollar sign before each number and carries the value three places to the right of the decimal place by creating the format "$.000."

NaOH Purchase Cost (Column D):

The purchase cost of the NaOH is simply calculated by multiplying the purchase price per liter (column C) by the number of liters per hour (column B). In **cell D5**, enter the formula

$$=B5*C5$$

and fill down to D24. Again, the numbers may be formatted to show a dollar sign and no numbers to the right of the decimal place and a comma when the number equals 1,000 or larger. Try creating a custom format to accomplish this.

NaOH Pumping Cost (Column E):

We have already learned that the cost of pumping the NaOH will depend on the NaOH concentration, and we've entered this cost relationship into the spreadsheet where we can see it. To use this relationship, enter the following formula into **cell E5**:

$$=IF(A5<13,I\$9*B5,I\$10*B5)$$

Translated, this means, IF the value in cell A5 (the NaOH concentration) is less than 13 mM enter into the cell E5 the product of the contents of cell I9 ($/L for C<13 mM) and the contents of cell B5 (L/hr), but IF NOT, then enter into cell E5 the product of I10 ($/L for C≥ 13 mM) and B5. We now Fill Down this column from E5 to E24 and set the format similar to that of column D.

Total Cost (Column F):

The last step is to simply sum the purchase cost (column D) and pumping cost (column E). In **cell F5**, enter the formula

$$=D5+E5$$

and fill down and format column F similar to columns D and E. Column F now contains the calculated total cost of the NaOH for the various concentrations of NaOH available.

The final version should look like Figure 6.4. This spreadsheet has several very useful advantages over the old hand-written calculation sheet. One of these is that a change in any of the given values can be easily accommodated. To illustrate, suppose we learn that the HCl flow rate really should have been 12100 *L/hr* instead of 11600 *L/hr*. By simply entering the corrected value of 12100 *L/hr* into cell I2, all of the calculations instantly adjust to give results based on this new value. Try this and see.

Another useful feature is the ability to print these important calculations on paper using a printer connected to the computer. Excel allows the user to print the whole sheet using the

"Print" command under the "File" menu. You may also print selected portions of the spreadsheet by highlighting the section to be printed (by dragging the mouse from the top left corner to the bottom right corner of that selection) and indicating in the print window of the "Print" command that only the "Selection" is to be printed. Alternately, one can set the print area by entering the formula for the area to be printed (e.g. A1:L24) in the "Print Area" box of the "Sheet" window under the "Page Setup" command (under the "File" menu).

Examination of the values in column F reveals that the total cost is a minimum value when the NaOH concentration is 25.0 *mM*. However, we can better see this relationship by plotting the values in column F against the values in column A. The next section will show us how to do that.

	A	B	C	D	E	F	G	H	I	J	K	L
1			NaOH Costs									
2	Conc.	Flow	Price	Cost	Pmp Cost	Tot Cost		HCl flow (L/hr) =	11600			
3	(mM)	(L/hr)	($/L)	($/hr)	($/hr)	($/hr)		HCl conc (mM) =	14			
4												
5	5.0	32,480	$.072	$2,345	$5,619	$7,964		NaOH cost ($/L)=				
6	7.5	21,653	$.114	$2,477	$3,746	$6,223		0.0058	* C^	1.4	+	.017
7	10.0	16,240	$.163	$2,642	$2,810	$5,452						
8	12.5	12,992	$.216	$2,808	$2,248	$5,055		Pumping cost ($/L)=				
9	15.0	10,827	$.274	$2,967	$2,100	$5,067		C<13mM	.173			
10	17.5	9,280	$.336	$3,117	$1,800	$4,918		C>=13mM	.194			
11	20.0	8,120	$.401	$3,260	$1,575	$4,835						
12	22.5	7,218	$.470	$3,395	$1,400	$4,796						
13	25.0	6,496	$.542	$3,524	$1,260	$4,784						
14	27.5	5,905	$.617	$3,646	$1,146	$4,792						
15	30.0	5,413	$.695	$3,764	$1,050	$4,814						
16	32.5	4,997	$.776	$3,876	$969	$4,845						
17	35.0	4,640	$.859	$3,984	$900	$4,884						
18	37.5	4,331	$.944	$4,088	$840	$4,928						
19	40.0	4,060	$1.032	$4,188	$788	$4,976						
20	42.5	3,821	$1.122	$4,286	$741	$5,027						
21	45.0	3,609	$1.214	$4,380	$700	$5,080						
22	47.5	3,419	$1.308	$4,471	$663	$5,134						
23	50.0	3,248	$1.404	$4,559	$630	$5,189						
24	52.5	3,093	$1.502	$4,645	$600	$5,245						

Figure 6.4. The final version of the calculation spreadsheet

Section 6.3   Graphing

Engineers must not only be competent in technical analysis, but they must also be effective salespeople. This is because engineering almost always requires that we convince others of the correctness of our analyses and decisions. Those other people include our colleagues who will be affected by the decisions made and our bosses who will provide the financial and personnel support to implement our decisions. The information we must communicate is often complex, so we must be able to convey that information quickly and effectively. One of the most valuable tools in that communication is graphical representation of data. This section will illustrate the usefulness of graphical representation by applying it to the acid neutralization problem.

The results that we calculated using a spreadsheet in the previous section led us to a conclusion about the optimum concentration of NaOH to purchase. By careful examination of column F of the spreadsheet we have created, we recognize that the total cost exhibits a minimum as a function of NaOH concentration, and we can identify the NaOH concentration where that minimum occurs. However, if we are trying to convince an audience concerning our decision, we would rather not force that audience to spend time examining a table of numbers. Instead, a graph of the total cost versus the concentration of purchased NaOH would help the audience quickly recognize that a minimum occurs at a certain concentration.

Now that we have decided to use a graph, the question at hand is how to generate that graph. One way would be to use a piece of graph paper and to plot the points by hand. A better method would be to use a spreadsheet program like Microsoft™ Excel, which has been developed with many sophisticated graphing features. For example, Figure 6.5 shows some types of graphs available in most versions of Microsoft™ Excel. Furthermore, such spreadsheets can produce graphs which are directly linked to the data in the spreadsheet, so that the graph automatically changes if some of the data are changed.

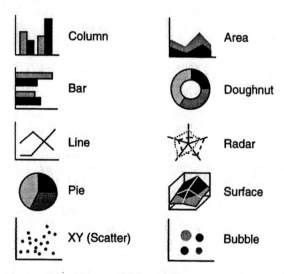

Figure 6.5 Some types of graphs available using most versions of Microsoft™ Excel

The procedure for creating a graph will vary from one program to another and can be learned from the tutorial or instruction material for the program being used. For illustration purposes, Table 6.4 describes the steps for constructing a graph using Excel™ in Microsoft Office™ 97 and 98.

Our objective is to plot the results of the calculations in order to arrive at an engineering decision about the optimum concentration of NaOH to use. As we noted in the previous section, the lowest total cost occurs at an NaOH concentration of 25.0 *mM*. The chart (Figure 6.6) makes it easier to see that minimum and also to see that it is not a sharp minimum, i.e. to see that any of the concentrations in the range of 20 to 30 *mM* produce roughly the same total cost. Therefore, we will try to purchase the 25.0 *mM* solution, but we will monitor other variables and will be willing to purchase any of the other concentrations in the 20-30 *mM* range if needed (e.g. if 25 *mM* becomes unavailable).

Table 6.4 Procedure For Creating a Graph using Excel™ in Microsoft Office™ 97 and 98.

1. <u>Highlight the two columns of data to be plotted.</u> If those two columns of numbers are contiguous (adjacent to each other), you can simply drag the mouse from the top of the left-hand column to the bottom of the right-hand column (while holding down the mouse button). If the columns are not contiguous, such as in the case of our spreadsheet where we want to plot column A against column F, first highlight (drag the mouse over) the numbers in one of the columns and release the mouse button and then hold down the "control" key (or the "apple" key for the Macintosh) on the keyboard while repeating the process for the second column of numbers.

2. <u>Open the ChartWizard.</u> This tool is found on the tool bar at the top of the window and looks like a bar (column) graph.

3. <u>Select the type of graph.</u> Clicking on the ChartWizard will open a scrollable window showing graph types similar to those shown in Figure 6.5. It is important to know that <u>only the XY Scatter type plots both the x-axis and the y-axis as scalable numbers</u>. The other types treat the x-axis as a series of non-quantitative entities (e.g. January, February, etc. or Western Region, Southern Region, etc.). Point the mouse at the type you want and press the mouse button once.

4. <u>Select the variation of graph type.</u> To the right of the scrollable window of graph types is a window that presents variations of the graph type you selected, such as representing the data with symbols only, with symbols and connecting lines, etc. Point the mouse at the variation you want and press the mouse button twice or press it once followed by clicking on the "Next" button.

5. <u>Verify the selection of the data.</u> A window will open showing the range of addresses for the data highlighted on the spreadsheet. Once you have edited the addresses and/or verified that they are correct, press the "Next" button.

6. <u>Customize your graph.</u> The third of four ChartWizard windows allows you to customize various features of the graph, including: Titles, Axes, Gridlines, Data Labels, and Legend preferences. The options for customizing each of these features can be accessed by placing the mouse on the desired tab at the top of the window and pressing the mouse button. Once you have chosen the preferences that fit your needs, press the "Next" button.

7. <u>Place your graph.</u> The final window allows you to specify where you want your graph to be located. You can choose to place the graph as an object in the spreadsheet that you are working in, or you can have Excel create a new spreadsheet and place the graph in that new sheet. Once you have made this selection, press the "Finish" button.

8. <u>Position your graph in the spreadsheet.</u> Once the graph has been completed, it can be moved around the spreadsheet. This is done by placing the cursor on the graph and, while holding down the mouse button, dragging the mouse to a new location. The move is completed by releasing the mouse button.

To understand the power of preparing the graph using a spreadsheet, you should recognize that the chart responds to the numerical values in the cells and will automatically change when those values are changed. For example, if the cost function needed to be updated or corrected,

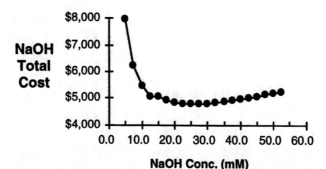

Figure 6.6 Chart of total cost versus NaOH concentration

simply correcting the values in cells H6, J6, and L6 would immediately result in the recalculation of the values in the other cells and the replotting of the chart to reflect those changes.

Let's conclude this chapter by adding the results of our calculations thus far to a stream table for the process flow diagram for our process. First, assuming that our dilute streams have the same density as water, we convert the flow rates and concentrations from our spreadsheets into mass flow rates for each chemical species.

HCl stream:
   HCl: 14 *mM* x 11,600 *L/hr* = 162 *gmol/hr* x 36.48 *g/gmol* = 5924 *g/hr* ≈ 6 *kg/hr*
   total stream: 11,600 *L/hr* x 1.00 *kg/L* = 11,600 *kg/hr*
   water: 11,600 - 6 *kg/hr* = 11,594 *kg/hr*

NaOH stream:
   NaOH: 25 *mM* x 6,496 *L/hr* = 162 *gmol/hr* x 40.00 *g/gmol* = 6,496 *g/hr* ≈ 6.5 *kg/hr*
   total stream: 6,496 *L/hr* x 1.00 *kg/L* = 6,496 *kg/hr*
   water: 6,496 - 6.5 *kg/hr* ≈ 6,490 *kg/hr*

Thus, the flowsheet and stream table are as presented in Figure 6.7.

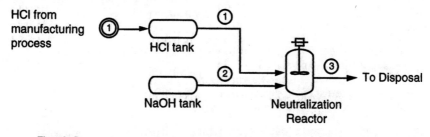

Flows kg/h

Line no.	1	2	3	
Stream	Acid	Base	Reactor	ABC Chemical Co.
Component	feed	feed	outlet	
HCl	6	—	—	Acid neutralization
NaOH	—	6.5	—	1×10^8 L/yr
H_2O	11594	6490	18096	Sheet no. 1
Total	11600	≈6496	18096	Dwg by Date
				Checked 1 Sep.1995

Figure 6.7 Process Flow Diagram with Stream Table

READING QUESTIONS:

1. Produce the spreadsheet in Figure 6.4 by following the steps outlined in the reading and printing the results on an available printer.

2. What was the advantage of entering the given information about the HCl concentration and flow rate, the NaOH price formula, and the NaOH pumping cost in their own cells rather than simply building that information into the formulas in other cells?

3. What was the advantage of using the formula =I$3*I$2/A5 rather than =I3*I2/A5 when preparing formulas for all the cells?

4. Produce the chart in Figure 6.6 by following the appropriate steps (e.g. those in Table 6.4).

HOMEWORK PROBLEMS:

1. Write the Excel expression for performing the calculation indicated by each of the following algebraic expressions:

    a. $[15.1 \tan(0.71)]^{4.3}$

    b. $\sqrt{\dfrac{A9 + G27}{C21}}$ where A9, G27, and C21 are cell addresses

    c. $\dfrac{21.3\, e^{D7}}{F19} + 3.85$ where D7 and F19 are cell addresses

2. Use a spreadsheet to estimate the accumulation in your retirement account during 20 years of your career as a chemical engineer. Use the following format and turn in a printout of the spreadsheet along with the answers to the questions posed below (this simple estimation is approximate, but is useful):

    a. Have the spreadsheet display the following as input values which can be changed:
       - a starting salary per year   put in $45000
       - an average annual raise (as a percentage)   put in 5%
       - projected interest earned per year (percentage)   put in 8%
       - fraction saved for retirement per year (as a percentage)   put in 6%

    b. Compute your salary for each year, which will be the previous year's salary plus the average raise.

    c. Compute the balance in your retirement account each year, which will be the previous year's balance plus interest earned on that balance plus the fraction of the current year's salary saved.

    d. How much is in your retirement account at the end of 20 years? How much does it change if you save 10% instead of 6%? There is a lesson to be learned from this problem, which is that you should save some of your income every year.

    e. How much is in your retirement account at the end of 20 years if you are very successful and rise to very responsible positions, causing your average raises to be 10% instead of 5%?

3. You will remember from your chemistry classes that under the appropriate conditions, gases obey the "Perfect Gas Law" or "Ideal Gas Law," namely

$$PV = nRT$$

Where P = pressure of the gas
               V = volume of the gas
               n = number of moles of the gas
               R = Universal Gas Constant (some useful values given in the front of the book)
               T = absolute temperature (e.g. in *Kelvin*) of the gas

Construct a spreadsheet to graph the volume of 1.0 *gmol* of a ideal gas at $0°C$ (273 *K*) as a function of the pressure. Your graph should show the volume in *Liters* on the vertical axis and pressure (for a range of 1 to 2 *atm* with at least ten intermediate values) on the horizontal axis. Turn in printed copies of the spreadsheet and the graph.

4. Construct an Excel spreadsheet to accomplish the following:

We wish to make cylindrical containers by drilling out aluminum cylinders

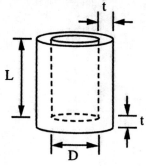

The constraints are:
     •The drilled-out volume must be 85 cm^3.
     •A minimum wall and bottom thickness of 0.4 *cm* must be allowed for strength.

Other information is:
     •The cost of aluminum is $.025/$cm^3$ (applied to the volume before drilling).
     •The drilling cost is $0.13*L + $0.34*D   for D < 3 *cm*
                         $0.13*L + $0.41*D   for D ≥ 3 *cm*
     where L and D are the length and diameter of the drilled hole (in *cm*), respectively

Objectives: Find the cost for various possible dimensions and find the minimum cost.

Construct your spreadsheet so that each numerical value given above is displayed in a cell on the spreadsheet and so that changing any one of those values will automatically change all calculations accordingly.

Also have Excel graph the total cost of the part as a function of the drilled diameter.

Turn in a printed copy of the spreadsheet and graph, both of these based on the given values. On the printed spreadsheet, write (by hand) the exact formula used for the bottom cell of each column of calculations.

Courtesy of Eastman Chemical Company, Longview, TX

CHAPTER 7

FLUID FLOW
(BRINGING THE BASE TO THE ACID)

Section 7.1   How Do Fluids Flow?

We have determined that we want to purchase some NaOH and add it to the HCl. Our next challenge is to provide a way to bring the NaOH in contact with the HCl. We probably will store the NaOH in a tank, and more NaOH will be added to the tank from time to time as we purchase additional amounts. But how do we bring the NaOH in contact with the HCl? The answer is that we will need to make the NaOH flow in a pipe to a location where that contact will occur. This chapter deals with some fundamental principles about making fluids flow and how those principles will help us with the problem at hand.

We usually group fluids into gases and liquids. **Gases** are characterized as loosely-associated molecules which are normally not close together and which travel through space for long distances (many times larger than the molecular diameter) before colliding with each other. The velocity of their travel depends on the temperature of the gas. **Liquids**, on the other hand, are characterized by molecules which are very close together (on the same order as their molecular diameter) and which are in collision with each other very frequently as they move around each other. The velocity of that motion and the rate of that collision depend on the temperature of the liquid. This picture of the molecular dynamics of gases and liquids supports the observation discussed in Chapter 4 that the densities of gases are much smaller than those of liquids.

In the acid-neutralization problem, the important fluids (NaOH and HCl solutions) are liquids. How do we move the NaOH solution at flow rates which can be adjusted and controlled in order to be appropriate for the flow rates of the HCl solution leaving our company's process? Is the task easier when the tank of NaOH is full than when it is only half full? To answer these questions and design an appropriate system for neutralizing the HCl, we need to introduce some principles of fluids and of fluid motion. As a foundation for that discussion, we need first to review the concept of "pressure."

The Concept of Pressure

Many of you have been introduced to the concept of "pressure" in a chemistry class where you learned about the ideal gas law which relates pressure, volume and temperature ($PV = nRT$). What is pressure and how does it relate to flow?

To begin answering these questions, let's first consider a particular volume of fluid which is surrounded by a boundary. As you may recall, a fluid consists of molecules which are free to move about to one extent or another, and can thus be distinguished from a solid where the molecular motion is much more restricted. The molecules which make up the fluid will exert a force on the surrounding boundary. The pressure of the fluid is defined as the total force (exerted on the boundary by the fluid molecules) divided by the surface area of the boundary it is acting upon. In other words, pressure is force per area.

In gases, pressure is caused by the kinetic energy of the molecules as they collide with the boundary and is related to the frequency and force of the collisions.   Increasing the temperature of a gas will increase the pressure, as predicted by the ideal gas law, by increasing the velocity of the gas molecules and thus the frequency and force of molecular collisions.   Similarly, decreasing the volume of a gas without changing the number of molecules will decrease the average distance between molecules and cause the frequency of molecular collisions to increase, which will again increase the pressure as predicted by the ideal gas law.

The situation is different for liquids where molecular movement is much more restricted because of the close proximity of the molecules. In fact, the molecules are so close that any attempt to squeeze them more tightly together causes them to push back in order to avoid getting any closer. This "pushing back" or repulsive force is the principal cause of pressure in liquids. Because this repulsive force is quite strong, pressure changes in liquids are accompanied by only small changes in volume.   Consequently, the number of molecules per volume changes very little with pressure for liquids, and we can assume that the density of a liquid is constant over a wide range of pressures. This is not true for the density of a gas which changes significantly with pressure.

We express pressure in two ways, *absolute* and *gauge* pressure, defined as follows:

Absolute Pressure: the total magnitude of force exerted per unit area
Gauge Pressure: the absolute pressure minus the prevailing atmospheric pressure

In other words,

$$\text{Gauge pressure} = \text{Absolute pressure - Atmospheric pressure} \qquad (7.1)$$

Pressure has units of force per area; common units for pressure are listed in Table 7.1.

Table 7.1 Some Common Units of Pressure

unit name	abbreviation	quantity in 1 atmosphere
pounds force per square inch	*psi*	14.7
Pascals	*Pa*	101,300
atmospheres	*atm*	1.0
millimeters of mercury	*mm Hg*	760

We usually designate whether a pressure is being expressed in absolute or gauge pressure by adding "(abs)" or "(gauge)" at the end of the units, e.g. 935 *mm Hg* (abs). In the special case of pressure expressed in *psi* (pounds force per square inch), the letter "a" or "g" is added to designate absolute or gauge, thus *psia* or *psig*.

Example 7.1

A man pumps up his automobile tire until the tire gauge reads 34.0 *psi*. If the atmospheric pressure in his community is 14.2 *psia*, what is the absolute pressure of the air in the tire?

Solution: The tire gauge reading is relative to the atmosphere, so the tire pressure is really 34.0 *psig*. From Equation 7.1

$$\text{Absolute pressure} \quad = \text{Gauge pressure + Atmospheric pressure}$$
$$= 34.0 \ psig + 14.2 \ psia = 48.2 \ psia$$

Up to this point in the section we have defined pressure and discussed the units needed to express the pressure quantitatively. How does pressure relate to flow? The answer to this question is that, in the absence of other forces, fluids tend to flow from regions of high pressure to regions where the pressure is lower. Therefore, pressure differences provide a driving force for fluid flow. This can be illustrated with a couple of simple examples from everyday life. When a tire is punctured, air flows out of the high pressure tire to the atmosphere, which is at low pressure. Pumping up your "super soaker" squirt gun pressurizes the liquid holding tank, and pulling the trigger opens a valve which allows the liquid to flow from the high pressure tank to the atmosphere. What other examples can you think of? All of these examples illustrate that pressure can be used to make fluids flow.

Shortly we will consider quantitative relationships between pressure and flow. However, let's first examine pressure variations in stagnant, or non-flowing, fluids.

Non-Flowing Fluids

We have suggested that our NaOH delivery system will contain a tank in which the NaOH will be stored. If a pipe is connected to the side of the tank near the bottom (Figure 7.1), our intuition tells us that NaOH will indeed flow out of the tank through the pipe. One of the factors which will determine the flow rate is the <u>pressure</u> at the bottom of the tank.

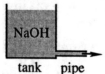

tank   pipe

Figure 7.1 A storage tank and exit pipe containing NaOH

Suppose that our tank filled with NaOH also has a valve preventing flow through the exit pipe (i.e. the fluid is stagnant). We can examine the values of the pressure at two different locations in the tank, say at the top and the bottom of the NaOH solution, which are at elevations z_1 and z_2, respectively (Figure 7.2).

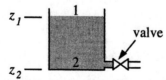

Figure 7.2 Defining the locations for comparison in a stagnant tank

The pressure at location "2" is higher than that at location "1" because of the weight of the fluid in the tank, and that difference in pressure is related to the height of the liquid in the tank. For fluids for which the density does not change very much with pressure (this is particularly true of <u>liquids</u>) that pressure difference can be calculated from the following equation:

$$P_2 - P_1 = \rho g(z_1 - z_2) \tag{7.2}$$

where ρ is the density of the fluid and g is the gravitational acceleration. Equation 7.2 is a well-known relation describing the pressure in <u>stagnant (non-flowing) liquids</u>. Note that the equation is not restricted to use at just the top and bottom of a tank, but is applicable to any two vertical positions within the fluid. The following example illustrates the use of Equation 7.2:

Example 7.2

For the tank depicted in Figure 7.2, if the NaOH solution is 8 *ft* (2.44 *m*) high, what is the pressure at the bottom of the tank? Assume that the density of the NaOH solution is the same as that of water (62.4 *lb$_m$/ft^3* = 1000 *kg/m^3*). Perform the calculation both in American units and metric units.

Solution: The strategy is to write in the values for the symbols in Equation 7.2

American units:

$$P_2 - P_1 = \left(62.4\frac{lb_m}{ft^3}\right)\left(32.2\frac{ft}{s^2}\right)(8ft) = 16,074\frac{lb_m}{ft\,s^2}$$

We have performed the calculation suggested by Equation 7.2, but the answer doesn't have units which we commonly use for pressure (e.g. *lb$_f$/in^2* or *psi*). The solution to this problem is to use the conversion factor which comes from the definition of a *pound-force* and which was introduced in Chapter 4, namely

$$\frac{32.2\ lb_m ft}{s^2 lb_f} \quad \text{or, inversely,} \quad \frac{1\ lb_f s^2}{32.2\ lb_m ft}$$

(see Example 4.2 for one illustration of how to use it). We can now use this conversion factor to express the pressure in more familiar units.

$$P_2 - P_1 = 16,074\frac{lb_m}{ft\,s^2}\left(\frac{1\ lb_f s^2}{32.2\ lb_m ft}\right)\left(\frac{1\ ft^2}{144\ in^2}\right) = 3.47\frac{lb_f}{in^2} = 3.47\,psi$$

Since $P_1 = 0$ *psig*, then $P_2 = 3.47$ *psig*

Metric units:

$$P_2 - P_1 = \left(1000\frac{kg}{m^3}\right)\left(9.80\frac{m}{s^2}\right)(2.44\ m) = 23,912\frac{kg}{m\,s^2}$$

Once again, we use the conversion factor from the definition of a *Newton*

$$\frac{1\ N}{1\ kg\,m/s^2} \quad \text{or, inversely,} \quad \frac{1\ kg\,m/s^2}{1\ N}$$

introduced in Chapter 4 (see Example 4.3 for an illustration of its use) and complete the calculation

$$P_2 - P_1 = 23,912\frac{kg}{m\,s^2}\left(\frac{1\ N}{1\ kg\,m/s^2}\right) = 23,912\frac{N}{m^2} = 23,912\ Pa$$

Again, $P_1 = 0$ *Pa* (gauge), so $P_2 = 23,912$ *Pa* (gauge)

Notice that this example illustrates that the pressure at the bottom of a column of liquid is greater than the pressure at the top, just as Equation 7.2 predicts. This is why a swimmer feels pressure on his/her ears at the bottom of a swimming pool. Since the pressure at the top of the liquid is atmospheric pressure (the same as the surrounding air), the pressure at the bottom is greater than atmospheric pressure. In fact, Equation 7.2 suggests that the pressure will increase linearly with depth.

Example 7.3

By what fraction will the pressure at the bottom of a tank decrease if the tank is drained to reduce the height of its liquid contents by 35%?

Solution: Since, for this case, location "1" is at the top of the liquid and is at atmospheric pressure, Equation 7.2 can be rewritten as

$$\text{Gauge Pressure at the Bottom} = \rho \bullet g \bullet \text{depth}$$

Therefore, decreasing the depth by 35% will reduce the pressure at the bottom by 35% as well.

From this discussion and the examples, you should recognize that Equation 7.2 is the important relation for problems involving stagnant liquids. For such problems, you should be able to use Equation 7.2 to determine an unknown pressure or height as required.

Principles of Fluid Motion

An important property of a fluid relating to its motion is its **mechanical energy**. For conditions of <u>steady state</u> and where <u>changes in density are not important</u>, the following equation applies:

$$\begin{matrix}\textit{Energy} \\ \textit{Gained}\end{matrix} = \left(\frac{P}{\rho} + \tfrac{1}{2}\alpha v_{avg}^2 + gz\right)_{downstream} - \left(\frac{P}{\rho} + \tfrac{1}{2}\alpha v_{avg}^2 + gz\right)_{upstream} = w_s - w_f \qquad (7.3)$$

The additive terms in the equation are all in units of energy per mass of fluid, but they represent different phenomena. To understand these terms, it's important to discuss how fluid flows.

In most cases of fluid flow of interest to chemical engineers, the fluid flows near a solid boundary (e.g. a pipe wall, wall of a column, etc.). In such situations, the velocity right next to the boundary can be considered to be zero (the fluid is stationary). In other words, the molecules of fluid closest to the wall "stick" to the wall. Further away, the fluid flows slowly, and still further away, the fluid flows at its maximum velocity. Hence, the velocity is not uniform (i.e. the value changes) across a fluid channel. For the common case of a circular pipe with a particular kind of flow called "laminar flow," we represent the varying velocity as shown in Figure 7.3, where the length of each arrow represents the magnitude of the velocity at that location. The envelope along the tips of the arrows describes the distribution of velocities, or the *velocity profile*.

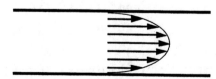

Figure 7.3 Velocities of a fluid in laminar flow through a circular pipe

Since the velocity varies with position within a flowing stream, it is convenient to speak of an average velocity, v_{avg}. The material balance which relates v_{avg} to the cross-sectional area of flow (A_{cs}) of a stream and the volumetric flow rate of that stream is

$$\dot{V} = v_{avg} A_{cs} \tag{7.4}$$

This relationship can be seen intuitively by imagining an element of fluid moving with a velocity v_{avg} through a cross-sectional area A_{cs} (Figure 7.4); in a unit of time, the fluid "sweeps" out a volume equal to the product given in Equation 7.4.

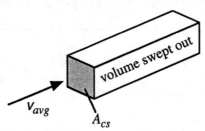

Figure 7.4 Graphical representation of Equation 7.4

Multiplying Equation 7.4 by the density of the fluid gives

$$\rho \dot{V} = \dot{m} = \rho v_{avg} A_{cs} \tag{7.5}$$

One implication of Equation 7.5 can be seen with the help of Figure 7.5 in which horizontal and downward-inclined sections of pipe are depicted, both carrying a liquid. Considering the horizontal pipe, a mass balance tells us that the mass flow rate into the pipe (location "1") must equal the mass flow rate out (location "2"), i.e. $\dot{m}_1 = \dot{m}_2$ or $\rho_1 v_{avg,1} A_{cs,1} = \rho_2 v_{avg,2} A_{cs,2}$. Furthermore, since a liquid is incompressible (constant density) and the pipe has a constant cross-sectional area, Equation 7.5 tells us that the average velocity must also be the same at "2" as at "1" ($v_{avg,1} = v_{avg,2}$). Considering the downward-inclined pipe, you might be tempted to think that the liquid should "accelerate" (flow faster) as it proceeds in the downhill direction, but all of the arguments made for the horizontal pipe apply here as well, and the average velocity at "2" must equal the velocity at "1." Therefore, you should see that <u>the average velocity of a liquid in steady flow through a constant-diameter pipe must be the same everywhere along the length of the pipe, even when the flow is uphill or downhill.</u>

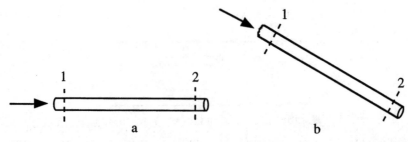

Figure 7.5 Flow of a liquid through horizontal (a) and downward-inclined (b) sections of pipe

Individually, the terms in Equation 7.3 relate to the following:

Kinetic energy: As a fluid moves from one location to another, that "translational" motion can be described by the average velocity. As taught in standard first-year physics classes, the energy associated with that motion is also the amount of energy (or "work") per volume of fluid needed to cause the fluid to accelerate from a stationary state to that average velocity and is

$$K.E. = \frac{1}{2}m(v^2)_{avg} \quad \text{or} \quad K.E.\ per\ mass = \frac{1}{2}(v^2)_{avg} \quad \text{(7.6a)}$$

where $(v^2)_{avg}$ represents the average value of the square of the velocity. However, v_{avg} is related to the flow rate and is more convenient than $(v^2)_{avg}$, and the two are not the same. To use v_{avg} in place of $(v^2)_{avg}$, we introduce a correction factor, α:

$$\text{(7.6b)}$$

$$K.E. = \frac{1}{2}\alpha m v_{avg}^2 \quad \text{or} \quad K.E.\ per\ mass = \frac{1}{2}\alpha v_{avg}^2$$

where α relates $(v_{avg})^2$, which is the same as v_{avg}^2, to $(v^2)_{avg}$. For many cases, α can be assumed to equal 1.0; you should make this assumption unless instructed to do otherwise.

Potential energy In the earth's gravitational field, an object or a fluid at any elevation can "fall" to a lower elevation and can give up energy (and do "work") in the process (for example, think of falling water turning a paddle wheel). Similarly, energy (or "work") is required to raise a fluid to a higher elevation. Again, first-year physics classes teach that the potential energy associated with a particular elevation is

$$P.E. = mgz \quad \text{or} \quad P.E.\ per\ mass = gz \quad \text{(7.7)}$$

Here, g is the gravitational acceleration and has units of acceleration (for example, cm/s^2), and the value of g was given in Table 4.3 in several selections of units. Also, z increases as we move upward in the vertical direction and is the height of the fluid relative to some defined (but arbitrary) reference height, so its value depends on the reference we select. This should not be a concern, however, since we will always be concerned with <u>changes</u> in energy, which will be independent of the reference height we select.

Energy associated with pressure (P/ρ): As will be shown in Chapter 10, the term P/ρ actually comes from a form of work called "flow work," but it is useful to think of it as a way of storing mechanical energy. As fluid flows, some of its kinetic or potential energy can be used to do work on the fluid in a way that increases the fluid pressure and decreases the kinetic or potential energy. The environment can also do work on the fluid to increase its pressure. In such cases, fluid pressure can be thought of as a way of storing energy in a flowing fluid. Similarly, that energy can be recovered as the fluid does work on itself to increase kinetic or potential energy or it can do work on its environment; in all these cases, the pressure decreases. Note that the units of pressure are force/area, which is the same as energy/volume. Therefore, P/ρ has units of energy per mass, although you will need to use the conversion factor from the definition of a *pound-force* to convert between these forms (see Example 7.2 and the discussion preceding it).

Work (w_s): This kind of work is called "**shaft work**" because it takes place through mechanical devices where a shaft is turned. The fluid EITHER <u>does work on its environment OR has work done on it by the environment</u>. Fluids do work on their environment through devices such as turbines. For example, large dams are usually built with large turbines which are turned by the water exiting at the bottom of the dam. The work done by the water as it turns the turbines decreases the energy of the water. Fluids have work done on them by their environment when pumps are used to increase the energy of those fluids. For example, pumps are used to lift water out of natural wells and to move it through piping systems toward various destinations, which increases the mechanical energy of the water. In Equation 7.3, the value of w_s will be positive when work is done on the fluid (e.g. by a pump) and will be negative when the fluid does work on its environment (e.g. in a turbine).

Friction (w_f):   The flow of a fluid produces **frictional effects** in the fluid as the fluid molecules flow past other fluid molecules or past the solid boundaries. These frictional effects decrease the mechanical energy of the fluid by converting it to heat. The term w_f represents the work which must be added to the system to overcome the friction, or alternatively, the energy which was lost to friction. The value of w_f is always positive.

A useful way to think of Equation 7.3 is that, for steady-state systems, the amount of energy gained by the fluid (per mass of fluid) equals the amount of work done on the fluid (per mass of fluid) minus the energy lost to friction. Equation 7.3 actually applies to any two locations in a continuous fluid stream. This leads us to a more general way of writing the **mechanical energy equation for steady-state incompressible flow**, which is

$$\left(\frac{P}{\rho} + \tfrac{1}{2}\alpha v_{avg}^2 + gz\right)_2 - \left(\frac{P}{\rho} + \tfrac{1}{2}\alpha v_{avg}^2 + gz\right)_1 = w_s - w_f \qquad (7.8a)$$

which can be re-written as

$$\frac{P_2 - P_1}{\rho} + \tfrac{1}{2}\left(\alpha_2 v_{2,avg}^2 - \alpha_1 v_{1,avg}^2\right) + g(z_2 - z_1) = w_s - w_f \qquad (7.8b)$$

where "1" and "2" represent any two reference points in the *same body of fluid*. Where flow is occurring, "1" and "2" are the inlet and outlet of a flow region, respectively. Furthermore, the terms w_s and w_f refer to the shaft work and friction which occur between locations "1" and "2."

Equation 7.8 is widely-used for determining the behavior of a fluid in steady-state flow of an incompressible fluid. As will be shown, it can be used to predict the pressure changes in a given flow system. Or it can be used to determine how much work a pump must provide to achieve a certain flow rate in a system. Or it can be helpful in designing a meter to measure the flow rate by monitoring the pressure variation in a pipe with varying diameter. Now let's apply this equation to some cases involving flowing fluids.

Special Case: No friction or shaft work

For the special case of flowing fluid where fluid friction can be considered negligible and where no shaft work is occurring, the mechanical energy equation (Equation 7.8) reduces to

$$\left(\frac{P}{\rho} + \tfrac{1}{2}\alpha v_{avg}^2 + gz\right)_2 - \left(\frac{P}{\rho} + \tfrac{1}{2}\alpha v_{avg}^2 + gz\right)_1 = 0 \qquad (7.9a)$$

or

$$\frac{P_2 - P_1}{\rho} + \tfrac{1}{2}\left(\alpha_2 v_{2,avg}^2 - \alpha_1 v_{1,avg}^2\right) + g(z_2 - z_1) = 0 \qquad (7.9b)$$

Equations 7.9a and 7.9b are forms of **Bernoulli's equation**, named after Daniel Bernoulli, who was an eighteenth-century scientist/mathematician who helped develop the concepts and solutions related to these equations.

Bernoulli's equation is useful because it helps us see that mechanical energy can be converted from one form to another. For example, let's consider the case where a tank of compressed gas at high pressure is opened, and gas exits the tank very rapidly. In this case, the energy associated with the pressure in the tank is converted to kinetic energy in the exiting gas stream.

•What conversions of energy are evident as water shoots vertically upward from a broken sprinkler, reaches a maximum height, and falls back downward?

•Can you think of other examples of this kind of fluid energy conversion?

In addition to qualitatively describing the conversion of mechanical energy from one form to another, Equations 7.9a and 7.9b comprise an important tool for quantitative calculations, as the following example illustrates:

Example 7.4

Octane (density = 44.1 lb_m/ft^3) flows downward in a tube at a velocity of 9.9 in/s. At a certain location, the pressure is known to be 1.60 $psig$. Just below that point, the diameter of the tube is reduced by one-half, and the velocity becomes 39.6 in/s. What is the pressure in the reduced section and 3 in below the original location where the pressure is known? Again, all values of α can be assumed to equal 1.0.

Solution:

Let's select the original point where the pressure is known as location "1" and the point where we want to know the pressure as location "2." The diagram for this system becomes:

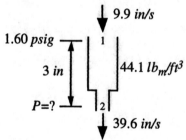

Rearranging Equation 7.10 and inserting the appropriate values:

$$P_2 - P_1 = \tfrac{1}{2}\rho\left(v_{1,avg}^2 - v_{2,avg}^2\right) + \rho g(z_1 - z_2)$$

$$= \tfrac{1}{2}\left(44.1\frac{lb_m}{ft^3}\right)\left[\left(9.9\frac{in}{s}\right)^2 - \left(39.6\frac{in}{s}\right)^2\right]\left(\frac{1\,lb_f s^2}{32.2\,lb_m ft}\right)\left(\frac{1\,ft}{12\,in}\right)^4$$

$$+ \left(44.1\frac{lb_m}{ft^3}\right)\left(32.2\frac{ft}{s^2}\right)(3\,in)\left(\frac{1\,lb_f s^2}{32.2\,lb_m ft}\right)\left(\frac{1\,ft}{12\,in}\right)^3$$

$$= -0.0486\frac{lb_f}{in^2} + 0.0766\frac{lb_f}{in^2} = 0.028\ psi$$

Therefore, $P_2 = P_1 + 0.028\ psi = 1.60\ psig + 0.028\ psi = 1.63\ psig$

Notice that all of the terms which were to be added (subtracted) had to be in the same units (lb_f /in^2), so conversion factors (including the conversion factor from the definition of a *pound-force* - see Example 7.2) were used for that purpose.

General Case: The effects of fluid friction

Example 7.4 not only illustrates how to solve problems with negligible friction and shaft work, it also provides an example of how to use the mechanical energy equation to solve problems in general. At this point, it would be useful to construct a formal list of recommended steps for using the mechanical energy equation. That list is provided in Table 7.2, with additional clarification from Figure 7.6 regarding some convenient reference locations.

Table 7.2 Steps for Using the Mechanical Energy Equation

1. Select the two reference locations "1" and "2." They should be locations where something is already known or where information is desired. For example, at the top of a liquid or where a fluid jet enters the atmosphere, the pressure is known to be atmospheric pressure, so those locations are good places to select the reference locations (Figure 7.6). Also, where velocities are small (like the top of the liquid in a large tank), the velocity can be assumed to be zero. If a particular location is where we want information (such as the pressure or velocity or height), that would also be a good place to select as a reference location, because it will then appear in the equation.

    When there is friction and/or shaft work, it is particularly important to assign "1" to the upstream or inlet side and "2" to the downstream or outlet side. If the direction of flow is not known, guess a direction for the flow. If the guess is wrong, the friction will turn out to be a negative number (which is impossible) or the work term will have the opposite sign from that which is expected.

    In all manipulations of the equation, keep careful track of the subscripts "1" and "2" to avoid confusion. Remember, that for each of the reference locations, the pressure, velocity, and elevation must apply to exactly the same location.

    Warning! The pressure in a liquid at the bottom of a tank cannot be determined by Equation 7.2 when the liquid is flowing, because that flow alters the pressure. Thus, the pressure at the entrance of a pipe at the bottom of a tank is usually unknown.

2. Determine the other velocities for expansions and contractions. Using a material balance such as Equation 7.5, the velocities at various diameters can be determined.

3. Eliminate terms which are known to equal zero. Depending on the details of the problem, a velocity, height, or pressure (gauge) could be considered zero.

4. Solve for the unknown value. The most useful method of proceeding is to solve for the unknown value algebraically (using the symbols of the other parameters) before inserting numbers for those known parameters. This often avoids doing unnecessary work, and the algebraic solution often provides additional insight.

Figure 7.6 Possible reference locations where the pressure equals atmospheric pressure

One of the most frequent applications of flowing fluids in chemical engineering is the transporting of fluids through closed cylindrical pipes. Using the mechanical energy equation, we can explore the role of pressure in such flow. From Equation 7.3, we see that a difference in pressure between two locations can be associated with flow between those locations. In fact, in a horizontal pipeline, the pressure difference is the driving force for flow and is necessary to overcome the friction associated with the flow. Let's illustrate that idea in Example 7.5:

Example 7.5

What is the relationship between the inlet and outlet pressures in a horizontal pipe of constant diameter?

Solution: First, we select the entrance and exit of the pipe as locations "1" and "2," respectively.

For a horizontal pipe, $z_1 = z_2$ (i.e. the potential energy does not change)
For constant diameter, $v_{1,avg} = v_{2,avg}$ (i.e. the kinetic energy does not change)
With no moving parts, $w_s = 0$

Thus, the mechanical energy equation becomes
$$\frac{P_2 - P_1}{\rho} = -w_f$$

Rearranging:
$$P_1 = P_2 + \rho w_f$$

Thus, we discover a principle of fluid motion:

In steady flow through a constant-diameter horizontal pipe, fluid flows through the pipe because the pressure is higher at the inlet end than at the outlet end, and, for constant fluid density, that difference in pressure divided by the density is equal to the friction produced by the flow.

The following example illustrates how to calculate the magnitude of the friction losses:

Example 7.6

In the case depicted in Example 7.5 where the fluid is water, what value of the friction per mass of fluid (w_f) is necessary to cause a decrease in pressure equal to
a) 10 psi (answer in Btu/lb_m)?
b) 68,900 Pa (answer in J/kg)? (remember, 1 $Pa = 1\ kg/m\ s^2$)

Solution: In this case, it is useful to use the first form of the equation from Example 7.5 but written as
$$w_f = \frac{P_1 - P_2}{\rho}$$

a) $w_f = \dfrac{10\,lb_f/in^2}{62.4\,lb_m/ft^3}\left(\dfrac{1\,Btu}{778.1\,ft\,lb_f}\right)\left(144\dfrac{in^2}{ft^2}\right) = 0.030\dfrac{Btu}{lb_m}$

b) $w_f = \dfrac{68,900\,kg/m\,s^2}{1000\,kg/m^3}\left(\dfrac{1\,J\,s^2}{1\,kg\,m^2}\right) = 68.9\dfrac{J}{kg}$

What causes friction in flowing fluid? Friction results from 1) fluid flowing past solid boundaries and 2) from fluid flowing past other fluid. Thus, we can expect that higher velocities, longer flow paths, and larger bounding surface areas (per volume of fluid) will produce greater overall friction, since all of these factors will increase fluid-solid interactions and fluid-fluid interactions. Since smaller pipe diameters result in higher velocities and higher surface areas per volume of fluid, a factor most influential in pipe flow is the pipe diameter; the smaller the diameter, the faster the pressure falls. Those who have done household plumbing know that 3/4-*inch* pipe allows for water to arrive at house fixtures at much higher pressure than does 1/2-*inch* pipe. The rate of pressure "drop" also depends on the flow rate; obviously, higher flow rates produce higher velocities and higher friction.

So far in this chapter, we have discussed the fact that we would store the purchased NaOH in a tank. We also established that the pressure at the bottom of a tank was greater than atmospheric pressure, and we learned how to calculate that pressure for a stagnant liquid. When the liquid is not stagnant but is flowing out of the tank, the pressure at the bottom of a tank where the liquid is exiting is not the same as in a stagnant tank. However, it is still greater than atmospheric pressure, and we have established that pressure differences cause flow through horizontal pipes. Thus, we could allow that pressure at the bottom of our NaOH tank to be the sole method of causing that solution to flow to where the HCl is. But there are some drawbacks to using this method of delivery in our process. Can you think of one or two drawbacks?

Pumps are used to compensate for friction

We just proposed one method of delivering the NaOH solution from the tank to the process, namely to rely on the higher pressure at the bottom of the NaOH tank to drive flow through the exit pipe from the tank. However, the pressure at the bottom depends on the liquid depth. Hence, the flow rate would also depend on how much NaOH solution is in the tank at the moment, and that level is expected to vary widely over time. We will probably fill the tank periodically as shipments arrive from the supplier (for example, by railroad car). Between shipments, the level in the tank will drop steadily. Therefore, the flow rate would tend to vary significantly unless we use some method of restricting that flow to offset these variations in level (such as constantly adjusting the exit valve).

A second drawback to using the pressure at the tank bottom to produce NaOH flow is that we could not achieve higher flow rates than produced by the current pressure at the bottom of the tank. This could be a problem, for example, if a high NaOH flow rate was needed while the tank was low, or if a higher flow rate was needed than could be provided even by a full tank.

It appears that we need a way to produce flow of the NaOH solution that doesn't depend on the level in the NaOH tank — we need a <u>pump</u>! We also need some pipe to convey the flow produced by the pump. Let's see how a pump affects the pressures in a pipe.

Example 7.7

What is the relationship between the inlet and outlet pressures in a horizontal pipe of constant diameter when a pump is providing flow?

<u>Solution</u>: Selecting the entrance and exit of the pipe as locations "1" and "2," respectively,

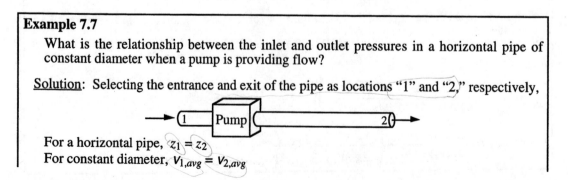

For a horizontal pipe, $z_1 = z_2$
For constant diameter, $v_{1,avg} = v_{2,avg}$

Now, the mechanical energy equation becomes

$$\frac{P_2 - P_1}{\rho} = w_s - w_f$$

Rearranging

$$P_2 = P_1 + \rho w_s - \rho w_f$$

Since $w_s = w_{pump}$,

$$P_2 = P_1 + \rho w_{pump} - \rho w_f$$

In the solution, we recognized that w_s in the equation is defined such that it has a positive value when work is done <u>on the fluid</u> by its environment, so w_s can be replaced by the work of the <u>pump on the fluid</u> (w_{pump}) since it represents the same thing.

The next section discusses some characteristics of the more common types of pumps and also discusses how the mechanical energy equation can be used to predict the performance when a pump is included in the system.

Section 7.2   Pumps and Turbines: Examples of Fluid Flow Devices

Pumps

Pumps are mechanical devices which move <u>liquids</u>. There are also devices (fans, blowers, and compressors) which move <u>gases</u>, but we won't talk about them here. Pumps move liquids by generating a high pressure at the pump outlet, which pushes the liquid into the outlet pipe. Pumps are usually grouped into two categories: centrifugal pumps and positive-displacement pumps, as discussed below.

Centrifugal pumps use the centrifugal force from a spinning disc-like "impeller" to produce liquid flow (Figure 7.7). The liquid enters the pump at 90° to the plane of the impeller and at the impeller center. The raised vanes on the impeller help to accelerate the liquid to the same speed as that of the impeller, which also forces the liquid to the collecting area around the periphery of the impeller and to the pump outlet. This type of pump can produce high flow rates, but because there is space between the impeller and the surrounding casing, this pump also allows significant liquid "slippage" (circulating flow inside the casing without entering the outlet). Thus, centrifugal pumps don't produce extremely-high pressures, and in fact, the outlet line can be closed completely without great danger to the pump or to other equipment.

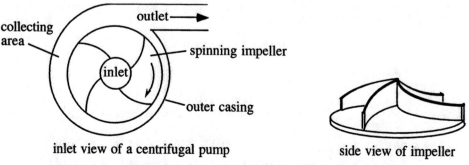

inlet view of a centrifugal pump   side view of impeller

Figure 7.7. Schematics of a typical centrifugal pump and impeller

Positive-displacement pumps come in many varieties, but they all rely on mechanical parts to directly push the liquid. For example, one type of positive-displacement pump employs a piston which moves inside a cylinder, while inlet and outlet valves control the direction of liquid movement through the pump (Figure 7.8a). Another type employs a screw on a turning shaft to push liquids through a cylinder (Figure 7.8b). A third example of positive-displacement pumps, one which produces low flow, is a peristaltic pump, which uses a roller to squeeze a section of flexible tubing which contains the liquid (Figure 7.8c). In contrast with centrifugal pumps, positive-displacement pumps can generate extremely-high pressures, and if the outlet line becomes closed, the pipe or other equipment in the outlet line can explode, causing great damage to equipment and, possibly, injury to people nearby. Thus, the use of positive-displacement pumps must also include the placement of "knock-out discs" or "relief valves" near the pump outlet to provide safe relief of pressure.

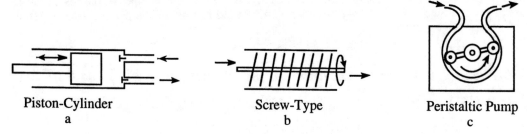

Piston-Cylinder   Screw-Type   Peristaltic Pump
    a   b   c

Figure 7.8. Some common types of positive-displacement pumps

You might have guessed that the subject of pumps was introduced at this time because the answer to our problem of controlling the NaOH flow rate includes the use of a pump. In other words, we should place a pump at the outlet of the NaOH tank and a pipe leading from the pump outlet to the reactor (Figure 7.9). It seems reasonable to assume that the pump will provide a flow rate which doesn't depend so much on the level of NaOH solution in the tank.

Let's take our problem a bit further — how does a pump provide flow through a pipe? We have already recognized that <u>a pump boosts the pressure of the liquid to some high value at the pump outlet</u>. Suppose that the tank in Figure 7.9 was open to the atmosphere, so that the liquid in the tank plus the effect of the pump would produce a pressure at the pump outlet that was significantly above atmospheric pressure. Suppose further that the pipe was open to the atmosphere at the exit end instead of entering the reactor. This means that the liquid leaving the pipe and entering the air would be at atmospheric pressure. In this case, the pressure of the liquid would decrease from some high value at the pump outlet to atmospheric pressure at the pipe exit! This situation is explored in Example 7.8.

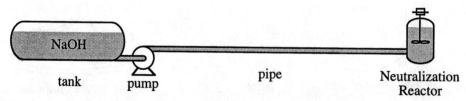

        tank   pump   pipe   Neutralization
                                                      Reactor

Figure 7.9. Using a pump to provide high flow rates independent of NaOH liquid level

---

Example 7.8

If the height of the NaOH solution in the tank is "H," how much work must the pump do to deliver the NaOH solution just to the reactor shown in Figure 7.9 (i.e. as though the exit end of the pipe was open to the atmosphere)?

<u>Solution:</u>   We'll select the top of the solution in the tank as location "1" and the pipe exit as location "2" (we'll do this because we know the pressures at these two locations).

For the open tank and pipe outlet, $P_1 = P_2 = P_{atm} = 0$ (gauge)
For a large tank, $v_{1,avg} \approx 0$
Thus, the mechanical energy equation becomes

$$gz_2 + \tfrac{1}{2}\alpha v_{2,avg}^2 - gz_1 = w_s - w_f$$

Recalling the definition of w_{pump} and rearranging

$$w_{pump} = w_s = w_f + \tfrac{1}{2}\alpha v_{2,avg}^2 - g(z_1 - z_2)$$

or

$$w_{pump} = w_f + \tfrac{1}{2}\alpha v_{2,avg}^2 - gH$$

The final solution of Example 7.8 provides an important insight into the need for a pump. Examination of each of the terms in that solution leads to the suggestion that the work of the pump is needed to overcome friction and to provide kinetic energy but that the pump requirements are offset by the available fluid height which assists with the flow, all of which agrees with our physical understanding. You should typically examine algebraic solutions such as the one above to determine whether the solution is consistent with physical arguments.

In Example 7.8, an expression was derived to calculate the "work" per mass of fluid. However, the desired work is often more conveniently expressed as work per time, or "power." Power can be found from the w_s term in the mechanical energy equation and the mass flow rate:

$$Power = mass\ of\ fluid\ per\ time \bullet work\ per\ mass\ of\ fluid = \dot{m}w_s = \rho\dot{V}w_s \qquad (7.10)$$

To complete our brief description of pumps, we should note that we have discussed and calculated the energy or work <u>delivered</u> by a pump to a liquid stream. But we sometimes need to know how much energy is actually needed to operate the pump, e.g. the electrical energy drawn by the motor of an electric pump or the fuel energy to run the internal combustion engine of an engine-driven pump. The power to operate the pump will be greater than the amount delivered to the fluid because of energy losses from such things as mechanical friction in the pump parts and fluid friction in the pumping chambers. We define the "efficiency" of a pump as

$$Efficiency = \frac{Power\ delivered\ to\ the\ fluid}{Power\ to\ operate\ the\ pump} \qquad (7.11)$$

The efficiency as defined in Equation 7.11 is always less than one for real systems. Typical values for the efficiency of a centrifugal pump range from 70 to 90%.

From the above description of the role of pumps, it seems that, for our problem of delivering NaOH and HCl to a reactor, pumps should be placed at the outlets of the tanks containing the NaOH and HCl as shown in Figure 7.10.

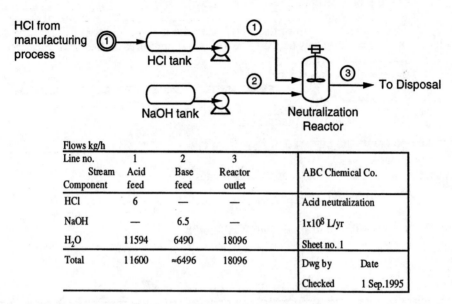

Flows kg/h					
Line no.	1	2	3	ABC Chemical Co.	
Stream	Acid	Base	Reactor		
Component	feed	feed	outlet		
HCl	6	—	—	Acid neutralization	
NaOH	—	6.5	—	1×10^8 L/yr	
H_2O	11594	6490	18096	Sheet no. 1	
Total	11600	≈6496	18096	Dwg by	Date
				Checked	1 Sep.1995

Figure 7.10 Process Flow Diagram for the acid neutralization process including pumps

Turbines

Turbines are like pumps operating in reverse in the sense that fluids do work on the turbine to rotate a shaft. In turn, the rotating shaft moves mechanical equipment to perform work or to turn a generator to produce electricity. Turbines vary greatly in design. For gases, turbines look most like propellers or multiple propellers in series. For liquids, turbines look more like centrifugal pumps operating in reverse or, in some cases, like paddle wheels. In all cases, the shaft work term w_s will have a negative value, and the work delivered by the fluid to the turbine, $w_{turbine}$, will equal $-w_s$.

Example 7.9

A dam holds a reservoir of water which drives a turbine-powered generator to provide hydroelectric power. Water flows from a pipe in the bottom of the dam (220 *ft* below the top of the water) through the turbine at a rate of 1650 *lbm/s* and then empties out of the turbine outlet pipe into the river below.

How much horsepower can the turbine theoretically produce? (The effect of friction and kinetic energy can be neglected in this case.)

Solution: We'll select the top of the reservoir and the turbine pipe outlet as locations "1" and "2," respectively (why did we choose those locations?). We will then solve the mechanical energy equation for w_s:

$$w_s = g(z_2 - z_1) = \left(\frac{32.2 \, ft}{s^2}\right)(-200 \, ft)\left(\frac{1 \, s^2 \, lb_f}{32.2 \, lb_m \, ft}\right)\left(\frac{1 \, hp \, s}{550 \, ft \, lb_f}\right) = -\frac{0.364 \, hp \, s}{lb_m}$$

$$w_{turbine} = -w_s = \frac{0.364 \, hp \, s}{lb_m}$$

$$Power = \left(\frac{0.364 \, hp \, s}{lb_m}\right)\left(\frac{1650 \, lb_m}{s}\right) = 600 \, hp$$

You will study the laws of fluid motion in more detail in your more advanced classes. Among the things you will learn in those classes is how to design piping systems, including how to predict the friction and associated pressure drop in pipes under various conditions and how to select pumps which provide the desired flow rates for piping systems. You will also learn how to determine the pressures and flow rates needed to obtain a given level of power from a turbine.

READING QUESTIONS:

1. In terms of molecular spacing, would you expect gases or liquids to be more compressible? Explain your answer.

2. For a stagnant liquid, how will the pressure at the bottom compare (greater or less) with the pressure at the top? In terms of physical forces, why do you think this is so?

3. Consider a liquid (constant density) pumped steadily along a pipe of constant diameter.

   a. We know from Chapter 5 that the mass flow rate into any section of the pipe must equal the mass flow rate out of that section. What does Equation 7.5 tell us about the outlet average velocity compared with the inlet average velocity? Is this answer different if the fluid is flowing uphill? downhill?

   b. If the pipe is horizontal, what happens to the pressure of the liquid as it is pumped through the pipe? What factor most strongly affects that phenomenon?

   c. If the liquid is pumped up a hill, what is happening to the potential energy of the liquid? Therefore, will the numerical value of the work term w_s in the mechanical energy equation be positive or negative? What does that imply about what the pump does to the energy of the liquid?

4. What were two drawbacks associated with the idea of using the pressure at the bottom of the NaOH tank to drive the flow of NaOH to where it will join the HCl?

5. Which kind of pump, a centrifugal pump or a positive-displacement pump, would you recommend for use in a home dishwasher or clothes washer? Support your answer.

HOMEWORK PROBLEMS:

1. A pressure-measuring device indicates that the blood pressure of a normal person oscillates between 120 *mm Hg* (at the peak of the heart beat) and 80 *mm Hg* (between heart beats). When the pressure-measuring device is taken off of the person and used to measure the pressure of the room air, it indicates a value of zero. For a normal person at sea level (where the air pressure = 1.0 *atm*),

   a. what are the <u>gauge</u> pressures at the peak and between beats?

   b. what are the <u>absolute</u> pressures at the peak and between beats?

2. If the pressure at a depth of 8 *ft* below the surface of a stagnant liquid is 3.5 *psi* above atmospheric pressure, what will the pressure of the liquid be at a depth of 48 *ft* below the surface? Hint: write the appropriate equation for both locations and take the ratio.

3. Two immiscible liquids sit in a closed tank as shown below.

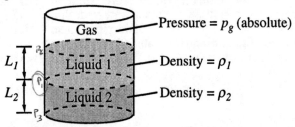

   The gas above the top liquid is at an absolute pressure of p_g. Furthermore, the densities and heights of liquid 1 and liquid 2 are ρ_1 and ρ_2, and L_1 and L_2, respectively. In terms of these given symbols, what is the absolute pressure at the bottom of the tank?

4. The left side of the mechanical energy equation (Equation 7.8a) has three terms: P/ρ, $\frac{1}{2}\alpha v^2$, and gz. Show that each of these terms has the dimensions of energy per mass of fluid by

   a. converting each term to units of Btu/lb_m, beginning with pressure in *psi*, density in lb_m/ft^3, velocity in *ft/s*, and height in *ft*.

   b. converting each term to units of J/kg, beginning with pressure in *Pa*, density in kg/m^3, velocity in *m/s*, and height in *m*.

5. One way to drain a liquid out of a tank or other container which has high walls is to use a "siphon." A tube is placed with one end in the liquid and the other end over the wall of the tank as shown.

   The tube is then filled with the liquid (usually by applying suction to the open end). As long as the elevation of the open end is lower than that of the top of the liquid, the liquid will flow out of the tank. Prove this by deriving an equation for the velocity coming out of the tube as a function of h. What happens when h approaches zero? (Assume that friction is negligible and that the top of the water in the tank is essentially stationary.)

6. A venturi meter is a device placed in a pipeline to allow for the measurement of flow through that pipeline. As shown in the figure below, the meter consists of a tapered reduction of the pipe diameter followed by an expansion back to the original diameter. Pressure taps in the upstream section and the narrow neck (called the "throat") of the meter make it possible to measure the difference between the pressures at those locations. The flow rate can be determined from that pressure difference using methods discussed in this chapter.

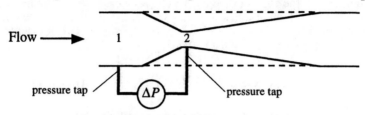

Venturi Meter for Measuring Flow Rate

As a summer engineer for the Valuchem Chemical Company, you are assigned to provide a way to monitor the flow rate of water (density = 62.4 lb_m/ft^3) through a 3-*in* diameter pipe in a process. You have ordered and installed a venturi meter with a throat diameter of 1.2 *in*. Now you need to relate the measured pressure difference to the flow rate.

a. Use a spreadsheet to predict the performance of the meter. Neglecting friction, have the spreadsheet calculate the pressure difference ($\Delta P = P_1 - P_2$) in units of *psi* for a series of flow rates ranging from 0 to 100 *gallons per minute (GPM)* in increments of 5 *GPM*. Have the pipe and throat diameters and the fluid density show in separate cells, so the values can be changed for different designs.

b. Have the spreadsheet plot a calibration curve that can be used with the flow meter, i.e. plot the flow rate (vertical axis) as a function of the pressure difference (horizontal axis).

c. The pressure indicator on the venturi meter shows a pressure difference of 2.75 *psi*. Using a pencil and ruler, draw the necessary lines on the graph in part b to estimate the flow rate of the water.

Turn in your derivation of the function(s) you used in the spreadsheet along with copies of the spreadsheet and the graph (show on the graph how you got your answer to part c).

7. A system consisting of a pump and pipeline is being designed to deliver water (density = 62.4 lb_m/ft^3) from a reservoir in the mountains down to the city 2800 *ft* below. The water must arrive at a water treatment plant in the city at a pressure of 450 *psig*.

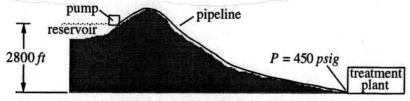

The flow rate of water is to be 63.5 *gal/s*, for which the friction in the pipeline is estimated to be 4.9 *hp s/lb_m*. How many horsepower must the pump deliver?

Hints: 1) Power is work/time and can be expressed in units of horsepower (*hp*) where 1 *hp* = 550 *ft lb_f/s*.

(Hints continued on the next page)

2) Assume that the kinetic energy term for flow in the pipe, e.g. at the entrance to the treatment plant, is small compared with the other energy terms.

3) Remember that each term in the mechanical energy equation has units of energy (or work) per mass of fluid, while power has units of energy (or work) per time.

8. A turbine operates by the impact of a high-velocity jet of steam against the turbine blades. The steam leaves the turbine at a lower velocity. For the turbine depicted below, steam from a high-pressure nozzle enters the air and strikes the blades of a turbine at an average velocity of v_i and leaves the turbine at an average velocity v_o.

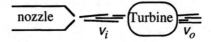

a. In terms of the given parameters and any others that may be needed, how much work can be generated by the turbine? Friction can be neglected.

b. If $v_i = 317$ *m/s* and $v_o = 126$ *m/s*, what is the amount of work (in *Joules* per mass of fluid) represented by the answer to part a)?

9. You have been asked to determine the power for a pump to deliver water ($\rho_{water} = 62.4$ *lb$_m$/ft^3*) over a hill which is 2500 *ft* high at its peak to a holding tank on the other side as shown below. The pipe outlet on the outlet side of the hill is a few feet below the peak of the hill, but that distance is unknown. The pipe is 6 *inches* in diameter (inside) and is the same diameter on both sides of the pump. The pressure of the water entering the pump inlet pipe is atmospheric pressure (0 *psig*).

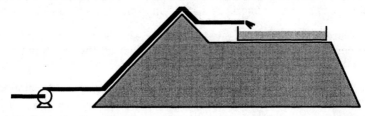

a. Draw a qualitative graph to show how the pressure changes as the fluid flows from the pump inlet to the point where it enters the tank.

b. If friction is neglected, where is the point of lowest pressure between the pump and the tank?

c. To avoid boiling (cavitation) in the pipe, the system has been designed so that the lowest pressure in the pipe is just at (but not below) the vapor pressure of the water (which is 17 *mm Hg* absolute). How much **power** must the pump deliver to the water in order to pump 150 *gal/min* over the hill? You may neglect friction for this calculation. If the pump efficiency is 78% and the cost of electrical power is $0.12 per *kW-hr* (i.e. $0.12 per 1000 *W* per hour of usage), how much will it **cost** to deliver the specified amount of water? (See hints #1 and #3 in problem 7 above.)

CHAPTER 8

MASS TRANSFER
(MIXING THE ACID AND BASE)

Now that we know how to deliver the NaOH to the HCl, we need to mix the NaOH and HCl together so that reaction can take place. More specifically, we need to bring molecules of NaOH into contact with molecules of HCl. This challenge falls into the area we call "mass transfer."

We will consider the movement of molecules move from one location to another by two mechanisms: **molecular diffusion** and **mass convection**. Following is a general description of both of these mechanisms and a discussion of how we can use this understanding to accomplish our task.

Molecular Diffusion

Molecular diffusion refers to the movement of individual molecules through a group of molecules without the aid of bulk fluid flow such as from stirring. For example, suppose a few molecules of liquid chemical "A" are placed in a particular location within a large group of molecules of liquid chemical "B" (Figure 8.1a). As all the molecules undergo random movement (in liquids, we call this Brownian motion), the molecules of "A" will eventually make their way to various locations throughout the available volume to produce a distribution like that shown in Figure 8.1b. This movement is the result of many random collisions between the molecules which are constantly in motion.

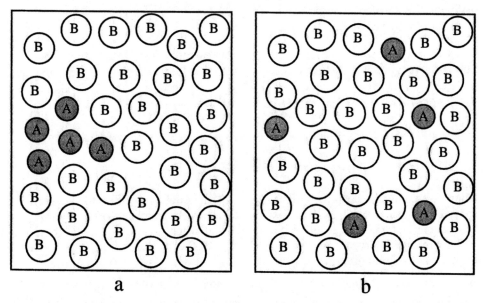

a   b

Figure 8.1. Illustration of molecular diffusion of liquid A (a) in liquid B leading to random distribution of the molecules (b)

The rate of transfer of "A" molecules through "B" molecules from one location to another by molecular diffusion is described mathematically by Ficks Law, named after Adolph Fick, a German physiologist who developed the equation to describe the transfer of materials in the human body. According to that equation, the rate of transfer of species A by molecular diffusion in the "x" direction (Figure 8.2) is:

where

$$\dot{N}_{A,x} = -D_{AB}A\frac{c_{A_2} - c_{A_1}}{x_2 - x_1} = -D_{AB}A\frac{\Delta c_A}{\Delta x} \tag{8.1}$$

$\dot{N}_{A,x}$ = diffusion transfer rate of species A (moles transferred per time, e.g. *gmol/s*, across area "A") in the "x" direction between locations "1" and "2"

A = cross-sectional area across which diffusion occurs (perpendicular to the x-direction)

D_{AB} = the binary diffusivity of species A in species B (a coefficient reflecting how easily "A" molecules move through "B" molecules), with units of area/time (e.g. *cm²/s*)

c_A = the concentration of species A (e.g. *gmol/L*)

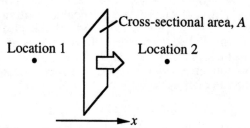

Figure 8.2 Depiction of molecular diffusion in the x-direction through area A

Equation 8.1 expresses the transfer rate in terms of a driving force (Δc_A), which tends to produce the diffusion, and a resistance ($\Delta x/D_{AB}A$), which tends to oppose that diffusion, so that

$$transfer\ rate = \frac{driving\ force}{resistance} \tag{8.2}$$

Compare this with the electrical analog provided by Ohms Law which can be written

$$current = \frac{voltage}{resistance} \quad OR \quad I = \frac{V}{R} \tag{8.3}$$

Equation 8.1 contains the binary diffusivity, D_{AB}, which describes the ease in which a molecule of species A moves through molecules of species B. In other words, when D_{AB} is large and transfer occurs rapidly, the resistance is small, and when D_{AB} is small and transfer occurs slowly, the resistance is large. As you would expect, D_{AB} depends on the properties of the molecules of A and B, including

•molecular size (which determines distances and spaces between molecules)

•molecular shape (including the presence of long chains which can tangle)

•molecular charge (which affects attractive or repulsive forces between the molecules)

As you might have also suspected, D_{AB} is really not a constant but varies with the physical conditions of the system. The most influential condition is the temperature, because that variable affects the motion of the molecules, and greater molecular motion makes it easier for molecules to move around each other.

Mass Convection

Mass convection is the method by which velocities and flow aid in the mixing of molecules of different types. Figure 8.3a depicts a small current of fluid moving in the direction represented by the dark arrow. That current carries molecules in its path to new locations, as represented in Figure 8.3b. Obviously, it is possible to have many small currents in a complex flow, causing rapid distribution of molecules. In addition, even in the presence of flow, molecules also continue to carry on molecular diffusion. However, when flow is present, the convection usually dominates as a mechanism of transfer.

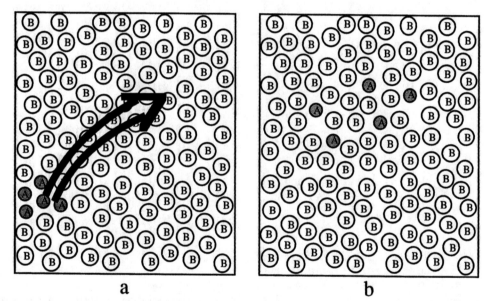

a   b

Figure 8.3   a) Illustration of mass convection showing a velocity current, represented by the dark arrow, (b) molecules which have been carried to new locations

You should recognize that mass convection distributes molecules much more quickly than does molecular diffusion alone. This is consistent with many experiences you have already had in life, such as when you stirred a liquid after adding an ingredient to speed up the distribution of that ingredient. In chemical engineering, it is often desirable to achieve maximum mixing of fluids, so chemical engineers sometimes design equipment to promote such mixing. That equipment may be in the form of small immovable baffles or vanes inside a section of pipe to produce many small currents of flow which collide and mix as the fluid flows through the pipe. Mixing tanks are also designed with propellers which are powered by a motor, and the design of the blades on the propellers varies according to the nature of the liquid being mixed.

Mass Transfer Through Phase Boundaries

A frequent application of mass transfer in chemical engineering processes is the transfer of mass across phase boundaries (interfaces), such as interfaces between a gas and a liquid, a liquid and a solid, or two immiscible liquids. Transfer across phase boundaries is a limiting factor in many chemical processes. How does a chemical engineer describe the rate at which such mass transfer occurs?

Describing mass transfer across phase boundaries is difficult because it involves the effects of both flowing fluids and molecular diffusion, as well as interactions at the interface. In this book, we will ignore any interactions at the interface and treat the transport in one direction only, perpendicular to the phase boundary. Figure 8.4 illustrates the physical situation of interest which consists of two phases separated by a boundary or interface. The direction of mass transfer is indicated by the large arrows. Let's suppose, because no flow is indicated in Phase I, that Phase I is stagnant and the transfer of species A in this phase is by diffusion as described by Equation 8.1. In contrast, the flowing fluid in Phase II influences the transport rate due to the effects of both convection and diffusion. Equation 8.1 does not apply to Phase II, and we need a simple way of describing transport across the phase boundary in the presence of flow.

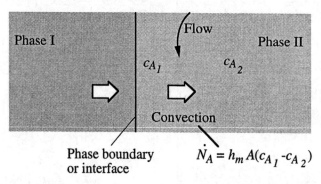

Figure 8.4 Mass transfer from a phase boundary by mass convection

To describe the complex effects of convection (including fluid motion, molecular diffusion, and interactions at the interface) on transport to or from phase boundaries, chemical engineers use a mass-transfer coefficient in the transport equation, as follows:

$$\dot{N}_A = h_m A(c_{A_1} - c_{A_2})$$ (8.4)

where

c_{A_1} = concentration of species A at the starting location ("1") of transfer (the interface)

c_{A_2} = concentration of species A at the ending location ("2") of transfer (the bulk of the fluid away from the interface)

$\dot{N}_A$ = convection transfer rate of species A (number of moles per time, e.g. *gmol/s*) through area "A" from location "1" to location "2"

h_m = mass-transfer coefficient, which accounts for the effects of diffusion and fluid motion (units of length per time, e.g. *cm/s*)

A = cross-sectional area through which the transfer takes place

Equation 8.4 is the appropriate expression for describing mass transfer whenever convection is present, such as in Phase II in Figure 8.3. Note that if convection had been present in Phase I as well, then Equation 8.4 would have been used for that phase also.

In Equation 8.4, the notion of a driving force is present in the concentration difference, just as it was in Equation 8.1 for diffusion. In the case of mass convection, the resistance is $1/h_m A$. In other words, when h_m is large and transfer occurs rapidly, the resistance is small, and when h_m is small and transfer occurs slowly, the resistance is large.

Multi-Step Mass Transfer

In many chemical processes involving mass transfer, the transfer actually occurs in several steps in series. Following are several examples.

Membrane Separation

One example of multi-step transfer is found in membrane separation, in which diffusion and convection occur in a process to separate chemicals using selective membranes, where the transferring substance passes through a membrane while other substances do not. The selectivity may be achieved on the basis of molecular size, as controlled by the size of the pores of the membrane (Figure 8.5). Other variations include membranes which select a material by solubility, i.e. will pass a material because it will associate with the material of the membrane, such as a hydrophilic (water-like) substance and a hydrophilic membrane, as opposed to a hydrophobic (oil-like) membrane. An example of membrane separation is the removal of urea from the blood of kidney patients using the membranes of a hemodialyzer.

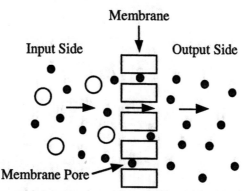

Figure 8.5 Schematic of a process of membrane separation

In membrane separation, the solution or mixture approaches the membrane on the input side, usually by convection. Similarly, the transferred material is carried away from the membrane on the output side, also by convection. While there may be fluid flow on both sides of the membrane, the "input" and "output" are defined in terms of the direction of transfer across the membrane. Further, the molecules of transferring material pass through the membrane pores by a process which is best described as diffusion through the molecular structure of the membrane.

Let's use this kind of separation to describe the relationships in multi-step mass transfer. The process involves at least three steps, as shown in Figure 8.6. These steps are:

1) transfer of the material from the bulk fluid on the input side of the membrane to the membrane surface (mass convection)

2) passage of the allowable material through the membrane (mass diffusion)

3) transfer of the filtered material, or filtrate, from the membrane surface on the output side of the membrane into the bulk fluid (mass convection)

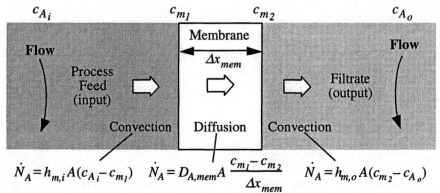

$$\dot{N}_A = h_{m,i} A (c_{A_i} - c_{m_1}) \qquad \dot{N}_A = D_{A,mem} A \frac{c_{m_1} - c_{m_2}}{\Delta x_{mem}} \qquad \dot{N}_A = h_{m,o} A (c_{m_2} - c_{A_o})$$

Figure 8.6 Close-up view of membrane separation showing the three parts of the process

Now, let's solve each of the transfer equations in terms of the concentration differences as follows:

$$c_{A_i} - c_{m_1} = \frac{\dot{N}_A}{h_{m,i} A}$$

$$c_{m_1} - c_{m_2} = \frac{\dot{N}_A \Delta x_{mem}}{D_{A,mem} A}$$

$$c_{m_2} - c_{A_o} = \frac{\dot{N}_A}{h_{m,o} A}$$

In a steady-state situation, the transfer of species A from the input side must equal the transfer through the membrane, which must also equal the transfer into the output side. In other words, $\dot{N}_A$ is the same for all three steps of the process. If the system is rectangular, as depicted in Figure 8.6, the cross-sectional area for transfer (A) is also the same for all three steps. We can now add these equations together and factor out $\dot{N}_A$ (notice that most of the concentration terms drop out), giving

$$c_{A_i} - c_{A_o} = \dot{N}_A \left(\frac{1}{h_{m,i} A} + \frac{\Delta x_{mem}}{D_{A,mem} A} + \frac{1}{h_{m,o} A} \right)$$

or

$$\dot{N}_A = \frac{c_{A_i} - c_{A_o}}{\dfrac{1}{h_{m,i} A} + \dfrac{\Delta x_{mem}}{D_{A,mem} A} + \dfrac{1}{h_{m,o} A}} = \frac{overall\ driving\ force}{\sum resistances} \qquad (8.5)$$

As we discussed earlier in the chapter, each one of the three steps of the process has a mass-transfer resistance, which reflects the fact that natural physical influences tend to inhibit or "resist" the transfer. For example, the transferring molecules may move very slowly through the fluid on the input side of the membrane, or may diffuse through the membrane very slowly, or may be slow to be transported away from the output side of the membrane. The mathematical terms representing these resistances are found to be added together in the denominator of Equation 8.5. The overall driving force is found in the numerator, and the total equation takes the same form as Equation 8.2.

Equation 8.5 helps us to see that the total resistance for the process is the sum of the resistances for the individual steps, so if one of those resistances is much larger than the other two resistances, that larger resistance will be dominant and will limit the total transfer rate. In such a case, we say that the larger resistance is the <u>limiting resistance</u>. Even if we reduce one or both of the other resistances (for example, by increasing the flow of bulk fluid), the mass-transfer rate will not increase very much. Mathematically, this is because the total resistance is dominated by the largest term in the sum and will not change very much if a small term is decreased. Physically, this is true because the overall transfer rate is limited or restricted by the slowest step. This concept of a limiting resistance is an important one which appears in many forms in the chemical engineering discipline. When an improvement is needed in a process, the chemical engineer will always want to know where the limiting resistance(s) is(are).

The use of Equation 8.5 and the concept of a limiting resistance are illustrated as follows:

Example 8.1

Liquid B flows on one side of a membrane, and liquid C flows along the other side of the membrane. Meanwhile, species A dissolved in both liquids transfers from liquid B through the membrane and into liquid C. The following data pertain:

concentration of A in liquid B	5.0 M
concentration of A in liquid C	0.1 M
thickness of the membrane	200 μm
diffusivity of species A in the membrane	1.0 x 10⁻⁹ m^2/s
area of membrane	1 m^2
mass-transfer coefficient on side B	7.0 x 10⁻⁴ m/s
mass-transfer coefficient on side C	3.0 x 10⁻⁴ m/s

a. What are the relevant mass-transfer mechanisms?

b. What is the transfer rate of A from the B side to the C side?

c. Calculate the limiting resistance.

d. Which of the following is most likely to increase the transfer rate?
   A) Increasing the flow rate of liquid B
   B) Decreasing the thickness of the membrane
   C) Decreasing the magnitude of the diffusivity
   D) Increasing the flow rate of liquid C

<u>Solution</u>: The drawing for this system is

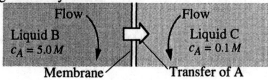

a. Because liquid is flowing on both sides of the membrane, the mass-transfer mechanism in those locations is mass convection. Through the membrane, species A transfers by molecular diffusion.

b. From Equation 8.5,

$$\dot{N}_A = \frac{c_{A_B} - c_{A_C}}{\dfrac{1}{h_{m,B}A} + \dfrac{\Delta x_{mem}}{D_{A,mem}A} + \dfrac{1}{h_{m,C}A}}$$

$$= \frac{5.0\,M - 0.1\,M}{\dfrac{1}{(7.0\text{x}10^{-4}\,m/s)(1m^2)} + \dfrac{2.00\text{x}10^{-4}\,m}{(1.0\text{x}10^{-9}\,m^2/s)(1m^2)} + \dfrac{1}{(3.0\text{x}10^{-4}\,m/s)(1m^2)}}$$

$$= \frac{5.0\,M - 0.1\,M}{1429\dfrac{s}{m^3} + 200000\dfrac{s}{m^3} + 3333\dfrac{s}{m^3}}\left(\frac{1000L}{m^3}\right) = 0.0239\frac{gmol}{s}$$

c. From the denominator of the above equation, the resistances can be seen to be

$$\frac{1}{h_{m,B}A} = 1429\frac{s}{m^3}$$

$$\frac{\Delta x_{mem}}{D_{A,mem}A} = 200000\frac{s}{m^3}$$

$$\frac{1}{h_{m,C}A} = 3333\frac{s}{m^3}$$

so, the membrane represents the limiting resistance, since it dominates the resistances so dramatically.

d. In view of the results of the previous parts of this problem, the options presented will have the following effects:

A) Increasing the flow rate of liquid B will decrease the resistance associated with mass convection on the B side, but since that resistance is so much smaller than the membrane resistance, it will have very little effect.

B) Decreasing the thickness of the membrane will likely have the largest effect, since the membrane resistance is limiting.

C) Decreasing the magnitude of the diffusivity will increase the membrane resistance and will therefore <u>decrease</u> the transfer rate.

D) Increasing the flow rate of liquid C will decrease the resistance associated with mass convection on the C side, but since that resistance is also much smaller than the membrane resistance, it will have very little effect.

In some circumstances, resistances in series are more equal (there is no clearly limiting resistance), but the largest resistance will still exert the largest effect on the transfer rate. This is illustrated in the following example.

Example 8.2

In patients with severe kidney disease, urea must be removed from the blood with a "hemodialyzer." In that device, the blood passes by special membranes through which urea can pass. A salt solution ("dialysate") flows on the other side of the membrane to collect the urea and to maintain the desired concentrations of vital salts in the blood. One geometry for hemodialyzer design is with flat membranes in a rectangular system. For such a geometry, consider the following typical values:

Blood side:
mass-transfer coefficient for the urea   0.0032 cm/s
average urea concentration within the dialyzer   0.020 gmol/L
Dialysate side:
mass-transfer coefficient for the urea   0.0015 cm/s
average urea concentration within the dialyzer   0.003 gmol/L
Membrane:
thickness   0.0016 cm
diffusivity of urea in the membrane   6.3 x 10⁻⁶ cm²/s
mass-transfer area   1.2 m²

a. Based on these values, what is the initial removal rate of urea? (Note: this rate will decrease as the urea concentration in the blood decreases.)

b. One might be tempted to try to increase the removal rate of urea by developing even better membranes for the hemodialyzer. Based on analysis of these characteristics, is such an effort justified?

Solution:

a. From Equation 8.5, the initial removal rate of the urea would be

$$\dot{N}_{urea} = \frac{c_{urea_{blood}} - c_{urea_{dialysate}}}{\dfrac{1}{h_{m,blood-side}A} + \dfrac{\Delta x_{mem}}{D_{urea,mem}A} + \dfrac{1}{h_{m,dialysate-side}A}}$$

$$= \frac{.02 - .003\, gmol/L}{\dfrac{1}{\left(.0032\dfrac{cm}{s}\right)(1.2\,m^2)} + \dfrac{.0016\,cm}{\left(6.3x10^{-6}\dfrac{cm^2}{s}\right)(1.2\,m^2)} + \dfrac{1}{\left(.0015\dfrac{cm}{s}\right)(1.2\,m^2)}}$$

$$\cdot\left(\frac{1000\,L}{m^3}\right)\left(\frac{1\,m}{100\,cm}\right)\left(\frac{60\,s}{min}\right) = 0.010\frac{gmol}{min}$$

b. The three resistances of importance are

$$\frac{1}{h_{m,blood-side}A} = \frac{1}{(.0032\,cm/s)(12000\,cm^2)} = 0.0260\,s/cm^3$$

$$\frac{\Delta x_{mem}}{D_{urea,mem}A} = \frac{.0016\,cm}{(6.3x10^{-6}\,cm^2/s)(12000\,cm^2)} = 0.0212\,s/cm^3$$

$$\frac{1}{h_{m,dialysate-side}A} = \frac{1}{(.0015\,cm/s)(12000\,cm^2)} = 0.0556\,s/cm^3$$

Improvement of the membranes would increase the transfer somewhat, since its resistance is significant relative to the other resistances. But the greatest potential to increase the removal rate is on the dialysate side, which has the largest resistance.

Up to this point, we have considered two types of "transfer mechanisms" or way in which mass transfer occurs, namely: 1) diffusion and 2) convection. As you solve mass transfer problems, one of the first things you will need to do is to identify the transfer mechanism. You will then be able to write down the expression required to solve the problem (Equation 8.1 or 8.4). The criterion for this decision has already been established in this chapter and is that, for the problems we will consider, if there is <u>no flow</u>, transfer will occur via <u>diffusion</u> (Equation 8.1) and if there is <u>flow</u>, transfer will take place via <u>convection</u> (Equation 8.4).

Chemical engineers use the principles of mass transfer in a number of types of systems such as the membrane system discussed above. Two additional systems, liquid-liquid extraction and gas-liquid absorption, are discussed below. In each case, transfer occurs in a series of steps, any one of which may be the limiting resistance to transfer. Try to identify which steps might be limiting in each of these systems.

Liquid-Liquid Extraction

When two liquids are mixed together, some pairs of liquids will blend to form a single phase while others will remain separate as two phases. For example, aqueous (water-based) liquids will not mix well with oil-based liquids. This can be seen when vegetable oil and water are combined to make a salad dressing; when shaken vigorously, the oil will distribute into small droplets within the water but will never completely mix with the water. Upon standing for a few minutes, the oil droplets will separate from the water again and will sit above the water in a separate phase. Liquids which will not mix with each other are said to be *immiscible*.

Given two immiscible liquids in contact with each other, a third chemical compound may be present in both phases but may preferentially distribute into one of the phases more than into the other. In other words, that compound may be more *soluble* in one of the phases than in the other. Going back to our example of the salad dressing, vinegar is also added to the dressing and distributes much more readily into the water phase than into the oil phase.

This difference in solubility can be used to create a driving force for mass transfer. For example, if one desires to remove the vinegar from an oil mixture, contacting the oil with water will cause most of the vinegar to leave the oil and transfer into the water. In chemical engineering, we say that the vinegar is being *extracted* from the oil. In general, liquid-liquid extraction is promoted when two immiscible liquids are brought into contact so that a particular compound will transfer from one phase to another. In that process, the transferring compound will travel through the original phase by diffusion or possibly convection, then through the liquid-liquid interface between them (i.e. the area where the phases are in contact), and then into the new phase by diffusion or convection (Figure 8.7). For example, gasoline spills result in the build up of certain organic compounds, like xylenes, in ground water, and one way to remove those organic compounds is by contacting the water with an organic phase (e.g. hexane).

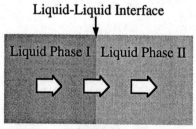

Figure 8.7 Illustration of mass transfer by liquid-liquid extraction

To cause liquid-liquid extraction to occur at a maximum rate, it is necessary to break the liquid phases into very small droplets so that the transferring substance does not have to travel very far to reach the interface. This is accomplished by having the separate liquids enter a vertical cylinder (called a *column*) through a *distributor* which breaks them up into droplets (Figure 8.8). The heavier liquid is introduced into the top of the column and the lighter liquid into the bottom so that gravity will cause them to flow through each other as the liquids attempt to trade places. To provide many small channels for the droplets to flow past each other, thousands of small solid shapes, called packing, fill the column (Figure 8.8). Many types of packing have been used, ranging from glass marbles to ceramic cylinders to stainless-steel rings, depending upon the flow characteristics and chemical reactivity of the liquids involved.

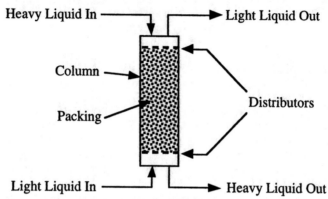

Figure 8.8 Schematic of a liquid-liquid extraction column

Gas-Liquid Absorption

Gas-liquid absorption is a simple variation of liquid-liquid extraction in which the two phases are a gas and a liquid instead of two liquids. The transferring substance travels through one of the phases (e.g. the gas) by diffusion or convection, then through the gas-liquid interface, and into the other phase (e.g. the liquid) by diffusion or convection (Figure 8.9). Everyday examples where this kind of transfer occurs include 1) bubbling carbon dioxide (from dry ice) through homemade root beer causing some of the CO_2 to be absorbed into the root beer solution, and 2) bubbling air through water in an aquarium causing oxygen to be absorbed into the water. In some cases, the rate of absorption into the liquid phase is increased by reacting the transferring species in the liquid to form a different compound so that the liquid-phase concentration of the transferring species remains low.

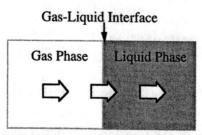

Figure 8.9 Illustration of mass transfer by gas-liquid absorption

A common application of gas-liquid absorption is for the removal of sulfur dioxide from combustion exhaust gas into a calcium carbonate or calcium hydroxide solution (which reacts with the sulfur compound). Removing the sulfur dioxide prevents its entry into the atmosphere where it can form sulfuric acid in the air and can produce "acid rain" which falls from the air onto the soil and ground water.

Conclusion

In each process we have discussed, a driving force tends to produce the desired transfer of a substance, and mass diffusion and/or convection provide a mechanism for that transfer to occur. Also, when transfer occurs through several regions or phases, one of the resistances can act as a limiting resistance. Many other kinds of mass-transfer processes could have also been described in this chapter, but all of them would have had the same basic characteristics.

This description of mass transfer and mixing has been to help us understand that we should provide some kind of mixing for the NaOH and HCl. Of course, some mixing will occur automatically if we simply bring the two streams together by joining the pipes from the HCl supply and the NaOH tank. The mixing will continue to take place as the combined stream continues along the pipeline. However, we probably will want to measure some chemical index telling us how good a job we have done in neutralizing the acid (we'll talk about this later), so we will want to have some point in the process where we know that the NaOH and HCl have been thoroughly mixed and fully reacted. Thus, we may want to augment the mixing by using a piece of equipment. We'll postpone deciding on what kind of equipment to use and how large to make it until the next chapter.

READING QUESTIONS:

1. Consider the molecular diffusion of species A through species B. How would each of the following affect the rate of that diffusion? In each case, explain your answer.
   a. The temperature of the mixture is increased.
   b. Species A is replaced with a species comprised of larger molecules.
   c. Species A is replaced with a species comprised of molecules which have long branches.

2. Consider the mass convection of species A from an interface into a flowing stream of species B as in Figure 8.4. How would each of the following affect the rate of that convection? In each case, explain your answer.
   a. The velocity of the flowing stream is increased.
   b. The interfacial area for transfer is increased.
   c. The concentration of species A at the surface of the interface is increased.

3. In each of the following, identify the driving force for mass transfer:
   a. A solid block of salt slowly dissolves in a stagnant pool of water.
   b. A quantity of a drug has been injected into the muscle of a patient so that it now slowly enters the circulation.

4. Molecules of a drug diffuse through a polymer membrane and into blood which flows next to the membrane.
    a. What kind of mass transfer (molecular diffusion or mass convection) occurs in the polymer phase and in the blood phase?
    b. If the mass transfer inside the polymer represents the limiting resistance, how would each of the following affect the overall rate of mass transfer? In each case, explain your answer.
        A. increase the flow rate of blood
        B. modify the drug to decrease the size of its molecules

HOMEWORK PROBLEMS:

1. In view of Equations 8.1 and 8.4, which describe molecular diffusion and mass convection, respectively, how would the rate of each type of transfer change if the difference in concentration of species A between the starting and ending locations were decreased by 40%?

2. Some discarded solid chemical waste dissolves slowly in a large drain pipe in which the water is stagnant. On a particular day, the dissolved chemical has a concentration of 0.16 M near the solid and is essentially zero at a location 13.6 m further along the pipe. The transfer rate (moles per time) of the chemical through the water in the pipe between those two points is 7.3 $gmol/min$.

    a. What equation describes this kind of transfer?

    b. Several days later, the chemical concentration near the solid has decreased to 0.105 M and is essentially zero at a location 9.9 m away. What will be the transfer rate between the point near the solid and the point 9.9 m away on this later day?

    c. On the next day, the heavy rains cause a current of water to flow through the drain pipe where the dissolving solid is located. The solid now dissolves twice as fast as on the previous day. If the concentrations are still 0.105 M near the solid and zero at the more remote locations, and if the cross-sectional area for transfer is 0.3 m^2, what is the value of the mass-transfer coefficient at this time?

3. The binary diffusivity of a particular salt in water at room temperature is 3.2×10^{-4} cm^2/s. That salt in solid form dissolves at the bottom of a beaker of water (8 cm high), while water flows across the top of the beaker at a high velocity, as shown.

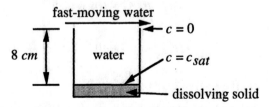

The concentration of the dissolved salt equals its saturation value (c_{sat}) at the top surface of the solid, and it equals zero at the top of the beaker (because of the high flow rate of the water flowing across). When the water inside the beaker is stirred at a particular rate, the solid dissolves at a rate 4 times greater than when the water in the beaker is not stirred (sits stationary). What is the value of the mass-transfer coefficient in the stirred condition (numerical value and units)?

4. The intermediate product of a chemical process is a chemical paste which is saturated with water. The paste is rolled into a thin layer and placed to dry in a flat area 30 *m* long by 15 *m* wide. Drying is enhanced by blowing hot air over the paste, and the mass-transfer coefficient is 0.017 *m/s*. Diffusion of water through the paste is rapid so that the concentration of water vapor at the surface of the paste remains constant at 0.002 *gmol/L*. Water must be removed from the paste at a minimum rate of 9.5 *L/min* (liquid volume). (Note: MW_{water} = 18 and ρ_{water} = 1 *g/cm³*)

   a. What mechanism controls the transfer of water from the paste into the air?

   b. What is the <u>molar</u> rate of water removal which corresponds to a <u>volumetric</u> removal rate of 9.5 *L/min*?

   c. What is the maximum concentration of water vapor allowable in the air (far away from the paste surface) if water must be removed at a rate of at least 9.5 *L/min*?

   d. Assuming that the mass-transfer coefficient remains constant, what can be done to increase the rate of water removal from the paste?

5. Your company manufactures hemodialyzers which have the characteristics described in Example 8.2. A colleague in the company has proposed replacing the membranes with better ones, which have the same thickness and area but for which the urea diffusivity in the membrane is 9.7 x 10⁻⁶ *cm²/s*. Assuming that the average concentrations of urea in the blood and dialysate are the same as with the old membranes, by what percentage would the new membranes increase the urea removal rate? In terms of resistances, explain why this turns out to be such a small improvement.

CHAPTER 9

REACTION ENGINEERING
(HOW FAST WILL THE REACTION GO?)

Section 9.1   Describing Reaction Rates

An important consideration in any chemical reaction, including the neutralization of HCl with NaOH, is that we provide sufficient time for the reaction to proceed. In the previous chapters, we developed a method to deliver the NaOH to the HCl and then discussed the need for the NaOH to mix well with the HCl. Presumably, the latter mixing would take place in a reactor of some sort. What kind of reactor would produce such mixing? Further, how large should that reactor be, and how much time should the reactants spend in the reactor so that the reaction proceeds to the extent desired? The question of reactor type and mixing will be addressed later in the chapter. To address the questions about reaction rate, let's begin our discussion by considering two additional questions:
   1) What physical variables affect the rate of a reaction between two chemical species?
   2) How do we describe the rate of a reaction?

Question #1: <u>What physical variables affect the rate of a reaction between two chemical species</u>, for example in the reaction $A + B \rightarrow C + D$? The most important variables include 1) the frequency at which the reacting molecules come in contact (collide) with each other, 2) the orientation and force of the collision, and 3) the energy requirements of the reaction. Now, let's discuss each one of these.

Frequency of molecular collision: The collision frequency depends on at least two variables. One of these is how many A and B molecules are present (i.e. their concentrations). This is obviously affected by whether the reaction takes place in a gas or a liquid phase, the temperature and pressure (especially for a gas), and the mole fractions of the reacting species. The second variable affecting the collision frequency is the velocity at which the molecules are moving. For gases, the molecules move long distances (compared with the molecular diameters) between collisions with other molecules. The velocity at which the molecules travel depends almost exclusively on the prevailing temperature. For liquids, the molecules are in contact with their neighbors as they bump and tumble around. Again, the "velocity" and frequency of contact depend upon the temperature of the liquid.

Orientation and force of the collision: While we cannot actually observe reactions taking place at the molecular level, we usually imagine that the reaction takes place as follows: Most of the collisions between A and B molecules produce no chance of reaction, either because the reactive parts of the molecules are not involved in the collision or because the molecules are not pressed together with enough force during the collision. However, in just a few of the collisions, one specific reactive part of an A molecule collides with a specific reactive location on a B molecule. In some of those cases, the force of the collision is sufficient to result in the formation of an "activated complex." An activated complex is a temporary molecule which is formed during the reaction sequence, which exists only for a very short period of time (e.g. 10^{-13} s), and then is transformed into the reaction product(s) by rearrangement of the atoms and/or by dissociation into multiple new compounds (as depicted in Figure 9.1).

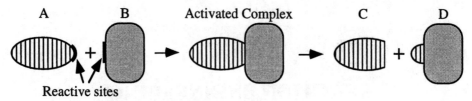

Figure 9.1 Depiction of the reaction $A + B \rightarrow C + D$

Energy requirements of the reaction: In the previous paragraph, we visualized <u>reactants</u> A and B colliding and forming <u>product</u> species C and D. It's important to note that we would also occasionally see the reaction <u>products</u> C and D collide to form the original <u>reactants</u> A and B (the reverse reaction). The observed rate of the forward reaction is actually the net difference of the many individual forward and reverse reactions. Furthermore, the relative numbers of the individual forward and reverse reactions depend upon how "favorable" the respective reactions are from an energy standpoint. In our simple model of a reaction, this might be like asking "With how much energy must A and B collide to produce a forward reaction? With how much energy must C and D collide to produce a reverse reaction?" We refer to the energy needed to produce a forward reaction as the "activation energy" (abbreviated E_a) for the forward reaction and, likewise, we refer to the energy needed to produce a reverse reaction as the "activation energy" for the reverse reaction. These energies are sometimes illustrated as shown in Figure 9.2. Notice that in Figure 9.2, the activation energy for the forward reaction is less than that for the reverse reaction and that the energy of the products is smaller than that for the reactants; in this case, then, the forward reaction is more favorable than the reverse reaction. However, as is discussed later, the actual rates of the forward and reverse reactions also depend on the concentrations of A, B, C, and D.

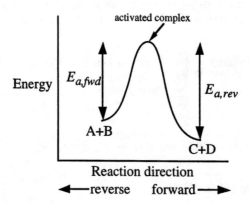

Figure 9.2 Representation of the activation energies for the forward and reverse directions

An important discovery related to the energy of reaction was that certain substances, called *catalysts*, lower the activation energies of some reactions, thereby allowing these reactions to proceed more readily (Figure 9.3). They do this by providing locations on the catalyst surface where the reactants not only bind to the surface but are altered or distorted in such a way as to make them more likely to react. Because these reactions proceed much more rapidly in the presence of catalysts, it is often possible to achieve desired reaction rates at much lower temperatures and pressures than would otherwise be required (thus, reducing the costs of building the equipment and operating the process).

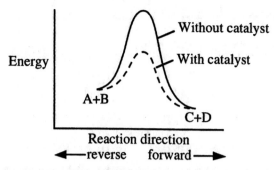

Figure 9.3 Representation of the influence of a catalyst on the energy of a reaction

As implied in Figures 9.2 and 9.3, every reaction can be described as having a forward direction and a reverse direction. For example, for the forward reaction A + B → C + D, the reverse direction would be C + D → A + B. When the rate of the reverse reaction is insignificant compared with that of the forward reaction, we describe the reaction as being "irreversible" and represent it by the forward reaction

$$A + B \rightarrow C + D$$

However, if the reverse reaction rate is significant, then we say that the reaction is "reversible," and it is more correct to write the reaction equation to indicate that the reaction can proceed in both directions, such as

$$A + B \leftrightarrow C + D \quad \text{or} \quad A + B \rightleftarrows C + D$$

In addition, the individual forward and reverse reactions may be influenced by different physical factors, and the net rate of the reaction will be influenced accordingly.

It's now time to address the second question posed at the beginning of the chapter.

Question #2: How do we describe the rate of a reaction? Let's suppose that we are considering the irreversible chemical reaction A + B → C + D. For liquid phases, we usually describe the rate of this reaction as shown in Equation 9.1.

$$reaction\ rate \left(\text{in units of } \frac{moles\ of\ A}{time\ volume} \right) = r_{reaction,A} = k_r c_A{}^n c_B{}^m \qquad (9.1)$$

As shown in Equation 9.1, the "rate" of the reaction is expressed in units of $moles/(time\ volume)$, where the volume is that of the reacting mixture (often the volume of the reactor vessel itself). Consistent with usual conventions in liquid-phase chemistry, the concentrations c_A and c_B are in units of moles per volume. For gas phases, the concentrations of the reacting species are often expressed as the partial pressure of the gas (in units of $atmospheres$), and the rate of the reaction is written

$$reaction\ rate \left(\text{in units of } \frac{moles\ of\ A}{time\ volume} \right) = r_{reaction,A} = k_r p_A{}^n p_B{}^m \qquad (9.2)$$

where the rate is in units of $moles/(time\ volume)$ or in units of $atmospheres/time$ (which can be converted into $moles/(time\ volume)$ using the ideal gas law) depending on the units of k_r.

The term "k_r" in Equations 9.1 and 9.2 is called the *reaction rate constant*, whose units depend on the form of the rate equation; for example, for rate expressions of the form shown in Equations 9.1 or 9.2, the units of k_r will depend on the values of n and m. Further, k_r (the reaction rate "constant") is indeed a constant at a given temperature, but its value changes with temperature. That dependence on temperature is usually expressed as

$$k_r = k_o e^{-E_a/RT} \tag{9.3}$$

where k_o is called the *frequency factor* (with the same units as k_r), E_a is the *activation energy* (in units of energy per mole) that we have discussed previously, R is the universal gas constant, and T is the temperature at which the reaction takes place (expressed as absolute temperature).

For a reaction which is accurately described by Equation 9.1 or 9.2, we say that the reaction is "n^{th}" order with respect to species A and "m^{th}" order with respect to species B. For example, if $n=2$ and $m=3$ for a particular reaction, the reaction would be described as "second order" with respect to species A and "third order" with respect to species B.

With some reactions, the order of the reaction with respect to each of the reactants matches the stoichiometry of the reaction equation. For example, for the reaction

$$2NO + O_2 \rightarrow 2NO_2$$

the reaction rate expression is

$$r_{reaction,NO} = k_r c_{NO}^2 c_{O_2}$$

where the order of the reaction with respect to NO is the same as the stoichiometric coefficient for NO (namely, 2), and the order of the reaction with respect to O_2 is the same as the stoichiometric coefficient for O_2 (namely, 1). However, in other cases, the stoichiometry in the reaction rate equation does not match the stoichiometry of the reaction. An example of such a reaction is

$$CO + \tfrac{1}{2}O_2 \rightarrow CO_2$$

for which the reaction rate expression has been found to be

$$r_{reaction,CO} = k_r c_{CO} \, c_{H_2O}^{0.5} \, c_{O_2}^{0.25}$$

which reflects the fact that the actual reaction mechanism is quite complex, including the involvement of water in several intermediate steps.

For the acid-neutralization reaction, the reaction of interest is

$$HCl + NaOH \rightarrow H_2O + NaCl \tag{9.4}$$

In this case, the reaction rate is first order in all reactants and can be expressed in units of *moles/(time volume)*, as

$$r_{reaction,HCl} = k_r c_{HCl} \, c_{NaOH} \tag{9.5}$$

For the reaction between HCl and NaOH (Equation 9.5), k_r has the units *volume/(moles time)*, e.g. *L/(gmol s)*.

The use of rate expressions such as those described above is illustrated in Example 9.1.

Example 9.1:

Ethylene glycol, a common antifreeze, is made from the reaction of ethylene chlorohydrin and sodium bicarbonate as shown below:

$$\begin{matrix} CH_2OH \\ | \\ CH_2Cl \end{matrix} + NaHCO_3 \longrightarrow \begin{matrix} CH_2OH \\ | \\ CH_2OH \end{matrix} + NaCl + CO_2$$

| ethylene chlorohydrin | sodium bicarbonate | | ethylene glycol | |

The reaction is essentially irreversible and is first order in each reactant, and the reaction rate constant at 82 °C is 5.2 *L/gmol hr* (from reference 1, p. 123).

A reaction mixture at 82°C with a volume of 17.5 *liters* contains ethylene chlorohydrin and sodium bicarbonate, both at concentrations of 0.8*M*. What is the production rate of ethylene glycol (in *gmol/hr*)?

<u>Solution</u>: The reaction rate expression is

$$r_{reaction,eth.chl.} = k_r c_{eth.chl.} \; c_{sod.bicarb.}$$

Inserting the given values:

$$r_{reaction,eth.chl.} = (5.2 \; L/gmol \; hr)(0.8 \; gmol/L)(0.8 \; gmol/L) = 3.33 \; gmol/L \; hr$$

$$production = (3.33 \; gmol/L \; hr)(17.5 \; L) = 58.2 \; gmol/hr$$

For the more complex case of reversible reactions we usually assign a reaction rate constant to each direction, such as k_r for the forward reaction and k_r' for the reverse reaction. For such cases, Equation 9.1 for liquid becomes

$$r_{reaction,A} = k_r c_A{}^n c_B{}^m - k_r' c_C{}^r c_D{}^s \tag{9.6}$$

which reflects the fact that the net rate of reaction is the difference between the actual rates of the forward and reverse reactions.

As we describe reactions, we should be aware that the reactants for a desired reaction often are also capable of undergoing an undesired, or side, reaction. In other words, we are often dealing with two competing reactions, such as

$$A + B \rightarrow C + D \tag{9.7a}$$

$$A + B \rightarrow E + F \tag{9.7b}$$

In such cases, the challenge is to maximize the desired reaction and minimize the undesired reaction. This may require selecting concentrations and temperatures that maximize the ratio of desired-to-undesired products. Also, as we write a mathematical equation describing the rate at which reactants A and B are consumed, we must be careful to specify whether we are referring only to the reaction in Equation 9.7a, the reaction in Equation 9.7b, or the total of both reactions.

The discussion in the previous paragraph has a direct relationship to the type of reactor chosen for a particular reaction. To illustrate, let's look at two idealized types of reactors, the continuously-stirred tank reactor (CSTR) and the plug-flow reactor (PFR) (Figure 9.4).

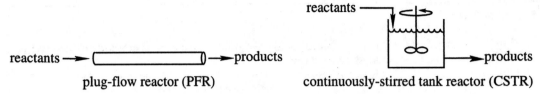

Figure 9.4 Illustrations of two idealized types of reactors

The PFR is a simple flow channel (e.g. a circular pipe) into which the reactants are fed. Each discrete volume of the reacting mixture which enters the reactor is assumed to flow through the reactor without mixing with the fluid upstream or downstream, i.e. to flow through as a "plug" — hence, "plug flow." The reaction proceeds as the mixture flows along the channel, so the reactant concentrations enter the reactor at the initial (highest) concentrations, decrease steadily as the mixture flows along the reactor, and exit at their final (lowest) concentrations. Similarly, the product concentrations increase as the mixture flows along the channel.

The idealized CSTR consists of a perfectly-stirred tank into which the reactants are fed. Because the tank is perfectly stirred, the concentrations in the tank are completely uniform and equal to the concentrations in the outlet. Furthermore, the concentration in the CSTR are the final concentrations, which are the lowest for the reactants (A and B) and the highest for the products (C and D). Under steady-state conditions, the concentrations are constant with time, which means that the rate the reactants are being consumed plus that rate that they leave the reactor adds up to exactly the rate that they enter the reactor (which is the same concept expressed by the steady-state material balances we constructed in Chapter 5, e.g. Equation 5.10).

We often select the type of reactor based on the kinetics of the desired reaction and the undesired side reactions. If the ratio of desired-to-undesired products is greatest when the concentrations of reactants (A and B) are high, we will probably select the PFR, since the reactants are at higher concentrations in the PFR. But if that ratio is highest when the concentrations of reactants are low, the CSTR is the reactor of choice, since the reactant concentrations are at their lowest values in the CSTR. That decision is illustrated in the following example:

Example 9.2

The following reactions are anticipated in a reactor:

$$\text{Desired reaction: } G + P \rightarrow X + W \qquad r_{reaction1,G} = k_{r1} c_G c_P^{1.2}$$
$$\text{Undesired reaction: } G + 2P \rightarrow Y + Z \qquad r_{reaction2,G} = k_{r2} c_G^{1.5} c_P^{2.5}$$

Which type of idealized reactor, PFR or CSTR, would maximize the ratio of the desired reaction versus the undesired reaction?

Solution: The ratio of desired-to-undesired reaction is

$$\frac{desired}{undesired} = \frac{r_{reaction1,G}}{r_{reaction2,G}} = \frac{k_{r1} c_G c_P^{1.2}}{k_{r2} c_G^{1.5} c_P^{2.5}} = \frac{k_{r1}}{k_{r2} c_G^{0.5} c_P^{1.3}}$$

Since this ratio increases with decreasing reactant concentration, we want to keep the reactant concentration as low as possible, which is better achieved with a CSTR.

Section 9.2   Designing the Reactor

Going back to the questions posed in the first paragraph of Section 9.1, we need to determine the volume of the chemical reactor for our problem. As indicated above, there are different types of reactors, and there are also many complex problems which could be considered. However, for the purposes of this introductory treatment, we will confine our discussion to continuously-stirred tank reactors (CSTR, see Figure 9.4) and irreversible reactions. We will then describe how to combine the reaction rate expressions in Section 9.1 with the material balances we used in Chapter 5 to solve this problem.

The $r_{reaction,A}$ term introduced in the previous section is the rate at which a reaction takes place, e.g. the rate at which the reactants are consumed, which is similar to the consumption term used in the material balances in Chapter 5. The $r_{consumption,A}$ term expresses the number of moles of A consumed per time in the reactor (Figure 9.5a), which accounts for the disappearance of A as it passes through the reactor and has units of moles/time. In contrast, $r_{reaction,A}$ expresses the number of moles of A consumed per time per unit volume of reaction mixture (Figure 9.5b) and has units of *moles/(time volume)* where the volume in the denominator refers to the volume of the reacting mixture. The relationship between the reaction rate and the consumption terms is

$$r_{consumption,A} = r_{reaction,A} V$$

(9.8)

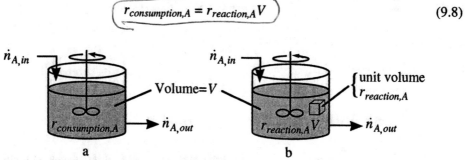

Figure 9.5 Illustration of the meanings of (a) $r_{consumption,A}$ and (b) $r_{reaction,A}$

Again, the volume in Equation 9.8 is the volume of the reacting mixture. Thus, if that mixture is a liquid which partially fills a reactor vessel (tank), the volume used in Equation 9.8 is the volume of the liquid, not the volume of the entire vessel. In the case of a gas mixture in a reactor vessel, the gas would fill the vessel, so the mixture volume would be the same as the volume of the vessel.

One of the unique features of a CSTR is that the contents are assumed to be perfectly mixed. Hence, as some of the mixture leaves the reactor in one or more outlet streams, the concentrations of species in those outlet streams are the same as the concentrations of those same species inside the reactor. Thus, since the concentrations used in the expression for $r_{reaction,A}$ (e.g. Equations 9.1 and 9.2) must be those in the reaction mixture (i.e., inside the reactor) and the concentrations inside the reactor are equal to the outlet concentrations for a CSTR, we can substitute for the concentrations in the reaction expression the outlet concentrations determined in our material balances (Equation 9.9).

$$r_{reaction,A} = k_r c_A^n c_B^m = k_r c_{A,out}^n c_{B,out}^m$$

(9.9)

There are a variety of problems we can solve using Equations 9.1(or 9.2) and 9.8 along with the material balances in Chapter 5, but we will focus on only one type of problem in this chapter:

determining the <u>reactor volume</u> needed to accomplish a desired conversion. To do this, for our introductory purposes, we recommend the following simplified procedure:

1. Perform the material-balance analysis for multiple species that we discussed in Section 5.2 (with a chemical reaction) to determine the values of relevant variables. In particular, the value of the outlet concentrations needed for the rate expression (e.g. Equation 9.1) will need to be determined, along with the value of the associated consumption/formation term(s).

2. Equate the consumption term to the rate equation multiplied by the volume (as in Equation 9.8) and solve for the reactor volume.

Now let's apply this procedure to some sample cases.

Example 9.3

Species A in liquid solution (concentration = 0.74 M) enters a CSTR at 18.3 L/s, where it is consumed by the irreversible reaction

$$A \rightarrow C \qquad \text{where} \qquad r_{reaction,A} = k_r c_A \ (k_r = 0.015/s \text{ and } c_A \text{ is in units of } gmol/L)$$

What reactor volume is needed so that the concentration of species A leaving the reactor equals 0.09 M? The density can be assumed to be constant.

<u>Solution</u>: (Note that the steps correspond to the instructions in Tables 5.1 and 5.2)

Drawing a diagram for this problem:

As per Table 5.2, we want to construct a <u>mole balance on A</u>. For this case (for a single input and single output stream), the mole balance becomes

$$\dot{n}_{A,in} + r_{formation,A} = \dot{n}_{A,out} + r_{consumption,A}$$

Species A is being consumed, but no species A is being formed, so $r_{formation,A} = 0$. This, along with substituting more convenient forms for the molar flow rates, gives

$$c_{A_{in}} \dot{V}_{in} = c_{A_{out}} \dot{V}_{out} + r_{consumption,A} \tag{a}$$

The value of the outgoing volumetric flow rate is not specifically given, so we need a <u>total mass balance</u>, which for a single input and single output stream, is:

$$\dot{m}_{in} = \dot{m}_{out}$$

which, in more convenient terms, is

$$\rho_{in} \dot{V}_{in} = \rho_{out} \dot{V}_{out}$$

Since the density is constant, this reduces to

$$\dot{V}_{in} = \dot{V}_{out} = \dot{V} \tag{b}$$

We can now calculate $r_{consumption,A}$ using Equations (a) and (b). Equation (a) becomes

$$r_{consumption,A} = c_{A_{in}} \dot{V}_{in} - c_{A_{out}} \dot{V}_{out} = (c_{A_{in}} - c_{A_{out}})\dot{V}$$

$$= \left(0.74\frac{gmol}{L} - 0.09\frac{gmol}{L}\right)\left(18.3\frac{L}{s}\right) = 11.9\frac{gmol}{s}$$

Up to now, everything we've done is a repeat of the material balances we learned in Chapter 5. The new step is to equate the $r_{consumption,A}$ term to the given rate expression times the reactor volume, where c_A (in the reactor) $= c_{A_{out}}$,

$$r_{consumption,A} = (k_r c_{A_{out}})V$$

or

$$V = \frac{r_{consumption,A}}{k_r c_{A_{out}}} = \frac{11.9\,gmol/s}{\left(\dfrac{0.015}{s}\right)\left(0.09\dfrac{gmol}{L}\right)} = 8{,}800\ L$$

Ideas to think about:

    1) what if $r_{reaction,A} = k_r c_A^2$?

    2) what if $r_{reaction,A} = k_r$?

As explained above, in determining the reactor volume, we need the concentrations of reacting species in the reactor outlet streams in order to use them for the reaction rate expression (Equation 9.9). In some cases, those outlet concentrations are not given and must be determined from material-balance calculations. Some reminders will be useful in preparing to make that calculation.

•Do not simply assume that the outlet concentration of a reactant will be a certain fraction, e.g. 25%, of the inlet concentration just because 75% the reactant is consumed in the reaction. This is because the volumetric flow rate of the outlet stream may be different from that of the inlet stream, due either to other streams entering the reactor or to significant changes in density. In such cases, a material balance will be needed to determine outlet concentrations.

•When formulating the mole balances for the reacting species, it is sometimes convenient to make the substitution into the mole balance based on Equations 4.5 and 5.9 that

$$\dot{n}_{A,out} = c_{A,out}\dot{V}_{out} \tag{9.10}$$

This will allow you to solve for the concentrations in the normal course of solving the equation set.

•If the outlet volumetric flow rate is not already known, it can be obtained through a total mass balance (remember, $\dot{m} = \rho\dot{V}$).

•If you already know the outlet molar flow rate of the important species ($\dot{n}_{A,out}$) and the outlet stream volumetric flow rate ($\dot{V}_{out}$), the concentration can be determined from a rearranged version of Equation 9.10, namely,

$$c_{A,out} = \frac{\dot{n}_{A,out}}{\dot{V}_{out}} \tag{9.11}$$

The use of these guidelines is illustrated in the following example:

Example 9.4

In the design of a process, separate liquid streams of pure species A and B will enter a CSTR, where they are consumed by the irreversible reaction:

$$2A + B \rightarrow C \quad \text{where} \quad r_{reaction,A} = k_r c_A \, c_B \quad (k_r = 24.7 \ ft^3/lbmol \ hr \text{ and } c_A$$
$$\text{and } c_B \text{ are in units of } lbmol/ft^3)$$

The molar flow rates of the inlet streams will be:

Species A: $\dot{n}_A = 110 \ lbmol/hr$ $MW = 59 \ lb_m/lbmol$

Species B: $\dot{n}_B = 68 \ lbmol/hr$ $MW = 133 \ lb_m/lbmol$

In the reactor, 90% of species A is to be reacted (i.e. 90% *conversion* of species A is desired), and the output stream will have a density of 50.5 lb_m/ft^3. What volume must the reactor have? Again, the density can be assumed to be constant.

Always draw a picture

Solution: The diagram for this problem is

A: 110 $lbmol/hr$, 59 $lb_m/lbmol$
B: 68 $lbmol/hr$, 133 $lb_m/lbmol$

$\begin{cases} 2A + B \rightarrow C \\ rate = (24.7 \ ft^3/lbmol \cdot hr) \ c_A \, c_B \\ 90\% \text{ conversion of } A \end{cases}$

$\longrightarrow$ 50.5 lb_m/ft^3, $\dot{V}_{out}$

reactor volume=V

Now that you're familiar with the approach, we'll abbreviate the description of the procedure.

Because the reaction rate expression contains the concentrations of two species, mole balances on both of those species will be needed.

- Mole balance on species A: (with substitution) $\dot{n}_{A_{in}} = c_{A_{out}} \dot{V}_{out} + r_{consumption,A}$ (a)

- Mole balance on species B: (with substitution) $\dot{n}_{B_{in}} = c_{B_{out}} \dot{V}_{out} + r_{consumption,B}$ (b)

- Total mass balance (to determine outlet flow): $\displaystyle\sum_{\substack{inlet \\ streams}} \dot{m}_{in} = \sum_{\substack{outlet \\ streams}} \dot{m}_{out}$

which, for this case, is $\dot{m}_{A_{in}} + \dot{m}_{B_{in}} = \dot{m}_{out}$

Substituting more convenient forms: $MW_A \dot{n}_{A_{in}} + MW_B \dot{n}_{B_{in}} = \rho_{out} \dot{V}_{out}$ (c)

- Additional Relationships:

Conversion: $r_{consumption,A} = 0.9 \, \dot{n}_{A_{in}}$ (d)

Stoichiometry: $\dfrac{r_{consumption,A}}{r_{consumption,B}} = \dfrac{2}{1} = 2$ (e)

This completes the equations we need to perform the material balance analysis. Now, we solve the equations together.

From Equation (c) $\dot{V}_{out} = \dfrac{MW_A \dot{n}_{A_{in}} + MW_B \dot{n}_{B_{in}}}{\rho_{out}}$

$$= \dfrac{\left(59 \dfrac{lb_m}{lbmol}\right)\left(110 \dfrac{lbmol}{hr}\right) + \left(133 \dfrac{lb_m}{lbmol}\right)\left(68 \dfrac{lbmol}{hr}\right)}{50.5 \dfrac{lb_m}{ft^3}} = 308 \dfrac{ft^3}{hr}$$

From Equation (d) $r_{consumption,A} = 0.9\left(110\dfrac{lbmol}{hr}\right) = 99\dfrac{lbmol}{hr}$

Combining Equations (d) and (e) gives

$r_{consumption,B} = (r_{consumption,A})/2 = (99\ lbmol/hr)/2 = 49.5\ lbmol/hr$

Therefore, from Equation (a),

$$c_{A_{out}} = \frac{\dot{n}_{A_{in}} - r_{consumption,A}}{\dot{V}_{out}} = \frac{110\,lbmol/hr - 99\,lbmol/hr}{308\,ft^3/hr} = 0.036\frac{lbmol}{ft^3}$$

and from Equation (b),

$$c_{B_{out}} = \frac{\dot{n}_{B_{in}} - r_{consumption,B}}{\dot{V}_{out}} = \frac{68\,lbmol/hr - 49.5\,lbmol/hr}{308\,ft^3/hr} = 0.060\frac{lbmol}{ft^3}$$

Finally, we equate $r_{consumption,A}$ to the given rate expression times the reactor volume,

$$r_{consumption,A} = (k_r c_{A_{out}} c_{B_{out}})V$$

giving $V = \dfrac{r_{consumption,A}}{k_r c_{A_{out}} c_{B_{out}}} = \dfrac{99\,lbmol/hr}{\left(24.7\dfrac{ft^3}{lbmol\ hr}\right)\left(0.036\dfrac{lbmol}{ft^3}\right)\left(0.060\dfrac{lbmol}{ft^3}\right)} = 1{,}860\ ft^3$

Now let's do the problem for the neutralization of HCl. The given information is:

$c_{HCl_{in}} = 0.014\ gmol/L,$ $\dot{V}_{HCl_{in}} = 11{,}600\ L/hr$

$c_{NaOH_{in}} = 0.025\ gmol/L,$ $\dot{V}_{NaOH_{in}} = 6{,}500\ L/hr$

The reaction is (see Equation 5.11)

$$HCl + NaOH \rightarrow NaCl + H_2O$$

and we will now add the information that the reaction rate equation is

$r_{reaction,HCl} = k_r c_{HCl}\, c_{NaOH}$ in units of moles of HCl or NaOH/(volume•time)

and we need the value of the reaction rate constant. The reaction rate constant is available for the reaction of HCl and NaOH at 25 °C and is[3]

$$r_{reaction,HCl} = 1.4 \times 10^{11}\ L/gmol\ s$$

Because values for the rate constant are not available for other temperatures (nor is an activation energy or a frequency factor available), we'll estimate the size of the reactor using the rate constant at 25°C. We will be more likely to actually operate our reactor at a higher temperature, such as the temperature that results from mixing the warm HCl and the stored NaOH; we'll show in the next chapter that the mixed temperature is approximately 50°C. At this higher temperature, the reaction rate will be higher than at 25°C (consistent with Equation 9.3), so our calculation of the reactor size based on 25°C will be a conservative estimate (the calculated reactor size will be larger than actually needed).

The next step is to determine the necessary reactor volume. We will assume a CSTR for this process. We make this assumption partly to keep our treatment within the scope of this introductory book. This assumption also seems justified for this design problem, because experience tells us that this is a fast reaction, and a CSTR should be adequate. Finally, we also use the CSTR to provide the mixing mentioned in Chapter 8.

To size the CSTR, we need to specify the final HCl concentration we wish to have coming out of the reactor. Using Utah state law as an example, that law specifies that the lowest pH allowable for water added to a natural lake or river is 6.5. We have actually calculated a NaOH concentration and flow rate that, if allowed to react fully, would result in a pH of 7.0 (complete neutrality). However, our reactor only needs to be large enough to provide sufficient reaction for satisfying the state law. Therefore, we need the concentration of HCl in the reactor to be

$$c_{HCl} = 10^{-6.5} = 3.16 \times 10^{-7} M$$

Now, we are ready to design the reactor. The working diagram is shown in Figure 9.6.

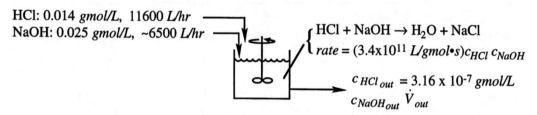

Figure 9.6 Working diagram to design the CSTR to react HCl with NaOH

Using the same approach used in Example 9.4, where the convenient forms of the molar flow rate and total mass flow rate are already substituted into the balances:

• **Mole balance on HCl:**

$$c_{HCl_{in}} \dot{V}_{HCl_{in}} = c_{HCl_{out}} \dot{V}_{out} + r_{consumption,HCl} \qquad (9.12)$$

• **Mole balance on NaOH:**

$$c_{NaOH_{in}} \dot{V}_{NaOH_{in}} = c_{NaOH_{out}} \dot{V}_{out} + r_{consumption,NaOH} \qquad (9.13)$$

• **Total mass balance** (with constant ρ):

$$\rho \, \dot{V}_{HCl_{in}} + \rho \dot{V}_{NaOH_{in}} = \rho \dot{V}_{out} \qquad (9.14)$$

• **Additional Relationships:**

Stoichiometry:

$$\frac{r_{consumption,HCl}}{r_{consumption,NaOH}} = \frac{1}{1} = 1 \qquad (9.15)$$

Molar flow rate balance: In Figure 9.6, the flow rate of the NaOH solution is given as an approximate value. This is because we recognize that the pH of the final solution is very sensitive to the balance between the HCl and the NaOH. Even the smallest variations in that balance will send the pH up or down beyond the environmentally-acceptable limits. Thus, we will vary the NaOH flow rate carefully as we monitor the pH (discussed in Chapter 12). For the sake of our present calculation, we recognize that we originally preset the inlet molar flow rate of NaOH to balance against the molar flow rate of HCl, as reflected in Equation 5.15, namely

$$c_{NaOH_{in}} \dot{V}_{NaOH_{in}} = c_{HCl_{in}} \dot{V}_{HCl_{in}} \qquad (9.16)$$

These equations are now solved together as follows:

From Equation 9.14 $\dot{V}_{out} = 11{,}600\ L/hr + 6{,}500\ L/hr = 18{,}100\ L/hr$

Using this result in Equation 9.12

$$r_{consumption,HCl} = c_{HCl_{in}}\dot{V}_{HCl_{in}} - c_{HCl_{out}}\dot{V}_{out}$$

$$= \left(0.014\frac{gmol}{L}\right)\left(11{,}600\frac{L}{hr}\right) - \left(3.16x10^{-7}\frac{gmol}{L}\right)\left(18{,}100\frac{L}{hr}\right) = 162\frac{gmol}{hr}$$

Substituting Equations 9.15 and 9.16 into Equation 9.13 and comparing with Equation 9.12 tells us that

$$c_{NaOH_{out}} = c_{HCl_{out}}$$

We are finally ready to equate the $r_{consumption,HCl}$ term with the reaction rate expression times the reactor volume, or

$$r_{consumption,HCl} = (k_r c_{HCl_{out}} c_{NaOH_{out}})V$$

giving

$$V = \frac{r_{consumption,HCl}}{k_r c_{HCl_{out}} c_{NaOH_{out}}}$$

$$= \frac{162\ gmol/hr}{\left(1.4x10^{11}\frac{L}{gmol\ s}\right)\left(3.16x10^{-7}\frac{gmol}{L}\right)\left(3.16x10^{-7}\frac{gmol}{L}\right)}\left(\frac{1\ hr}{3600\ s}\right) = 3.22\ L$$

Clearly, this represents a very small reactor, especially considering that the flow rates of the reacting streams combine to give 18,100 $L/hr = 5\ L/s$. Also, remember that at $50\,°C$, the reactor will proceed even more rapidly, and the required reactor size will be smaller still. This calculation is simply telling us that the reaction proceeds so rapidly that only a very small residence time in the reactor is needed to achieve the desired results. In fact, the mixing together of the two streams in a pipe will provide the needed volume and time.

Our conclusion from this design exercise is that a sophisticated reactor will not be necessary. Instead, the reactor function will be provided by the mixing point in the pipe system where the HCl and the NaOH streams join together. Some provision should be made at that junction to ensure rapid mixing so that the reaction is not hindered by the lack of contact between HCl and NaOH molecules, as discussed in the previous chapter. While adequate mixing may occur simply as a result of the flow patterns at the junction, some device to enhance that mixing, such as a simple in-line mixer, may be advisable. The process flow diagram for our process would now look like Figure 9.7.

References

1. Levenspiel, O., *Chemical Reaction Engineering*, 2nd ed., NY: John Wiley & Sons, Inc., 1972.

2. Hougen, O., and K. Watson, *Chemical Process Principles, Part 3, Kinetics and Catalysis*, NY: John Wiley & Sons., Inc., 1947.

3. Laidler, K.J., *Chemical Kinetics*, 3rd ed., NY: Harper & Row, 1987.

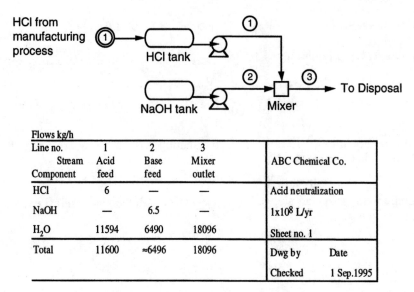

Figure 9.7. Process flow diagram for the acid neutralization process with an in-line mixer

The flows table within the figure:

Flows kg/h Line no.	1	2	3	ABC Chemical Co.
Stream	Acid	Base	Mixer	
Component	feed	feed	outlet	
HCl	6	—	—	Acid neutralization
NaOH	—	6.5	—	1×10^8 L/yr
H₂O	11594	6490	18096	Sheet no. 1
Total	11600	≈6496	18096	Dwg by Date
				Checked 1 Sep.1995

READING QUESTIONS:

1. Consider the gas-phase reaction

$$A + B \rightarrow C$$

   How would each of the following affect the rate of that reaction? Explain each answer.
   a. The temperature of the reaction is increased.
   b. The pressure of the reaction is increased.
   c. A catalyst specific to this reaction is added.

2. In your company, a particular reaction is produced in a CSTR. The design group has come to you for advice concerning the building of a new plant using this reaction. They propose to use a reactor with only half the volume as the reactor in the original plant and still produce the same amount of product. Furthermore, the concentrations will probably be the same as in the original plant. What kind of change could you propose that would make this possible? Why would this work?

3. The following two liquid reactions are known to occur simultaneously in a reactor:

$$A + B \rightarrow C + D \qquad r_{reaction, A,1} = k_{r1} c_A^5 c_B^3$$

$$A + B \rightarrow E + F \qquad r_{reaction, A,2} = k_{r2} c_A^2 c_B$$

   If E and F are preferred as products over C and D, which type of reactor (PFR or CSTR) is desirable? Why?

4. Suppose that the acid neutralization process is carried out by joining together the pipelines from the acid and base tanks without the recommended in-line mixer. Suppose further that pH meters were positioned at two locations, 3 *cm* and 20 *cm* downstream from the junction of the pipelines and that the pH 3 *cm* from the junction fluctuated much more than the pH 20 *cm* from the junction. What would these measurements indicate concerning the value of an in-line mixer? Explain your answer.

HOMEWORK PROBLEMS:

1. Show that for a rate equation of the form given in Equation 9.1, the units of k_r are

$$(moles)^{1-n-m} \ (volume)^{n+m-1} \ (time)^{-1}$$

where n and m are the exponents of the concentrations as shown in that equation.

2. Show that for a rate equation of the form given in Equation 9.2 with the rate in units of pressure/time, the units of k_r are

$$(pressure)^{1-n-m} \ (time)^{-1}$$

where "pressure" refers to any of the standard ways of expressing pressure, e.g. *atmospheres* and where n and m are the exponents of the partial pressures as shown in that equation.

3. A reaction vessel is filled with a solution at 22.9°C containing sulfuric acid and diethyl sulfate. The following reaction is first order in each reactant and occurs with an irreversible reaction rate constant of 6.74 x 10^{-4} *L/gmol s* (from reference 1, p.88):

$$H_2SO_4 + (C_2H_5)_2SO_4 \rightarrow 2 \ C_2H_5SO_4H$$

sulfuric   diethyl
acid   sulfate

When the sulfuric acid concentration is 0.53M, and the diethyl sulfate is at 0.28M, what reaction rate will be observed?

4. The reduction of nitrous oxide gas, an environmentally-important reaction, is:

$$2 \ NO \rightarrow N_2 + O_2$$

At 1620 K, for a reaction rate expressed in *gmol/L s* and the amount of NO expressed in *atmospheres*, the reaction rate constant for this irreversible reaction is 0.0108 *gmol/L s (atm)²* (from reference 2, p. 813).

   a. Assuming that NO is the only reactant, use the units of the rate constant to determine the order of this reaction in terms of NO.

   b. If a reactor is designed to reduce NO at a rate of 0.056 *gmol/min L* at 1620 K, what partial pressure of NO is needed in the reactor?

5. Butadiene sulfone (MW=118) can be produced by the following irreversible liquid-phase reaction at 190°F and 160 *psia*:

$$Butadiene + SO_2 \rightarrow Butadiene \ Sulfone$$

Pure SO_2 is fed to the reactor in one stream. Pure butadiene (MW=54) is fed separately into the reactor at a molar flow rate which is 25% more than that required to react with all of the SO_2, and 70% of the entering butadiene is converted into product. The density of the stream leaving the reactor is 42.2 *lb_m/ft^3*.

   a. For an SO_2 flow rate of 100 *lbmol/hr*, determine the volume of the CSTR required to achieve the specified 70% conversion. The rate of reaction is described by

$$r_{reaction, butadiene} = k_r c_{butadiene} c_{SO_2}$$

   where $k_r = 4.44 \ ft^3/lbmol \ hr$.

   b. Would the addition of an inert (non-reacting) liquid to the butadiene feed increase or decrease the required reactor size? Why?

6. As an engineer in a production facility, your assignment is to specify the size of a reactor needed to react a liquid stream (33 *L/min*) containing species G (concentration = 0.19 M). The goal is to produce a reactor outlet stream with a concentration of G equal to 0.04 M. To accomplish that, a second stream containing species J (concentration = 1.3 M) is also to enter the reactor but at 75% of the volumetric flow rate of the first stream, as shown.

$$\dot{V} = 33 \ L/min, \ c_G = 0.19 \ M$$
$$c_J = 1.3 \ M$$
$$\dot{V} =$$

Volume = ?

$$c_G = 0.04 \ M$$

The irreversible reaction is

$$G + J \rightarrow P$$

where the reaction rate only depends on species G according to the following kinetic relation:

$$r_{reaction,G} = \left(1.8 \ \frac{L}{gmol \ min}\right) c_G^2$$

Given these requirements, what size reactor (L) is needed to produce these results? (Assume equal densities for all streams)

7. For the acid-neutralization process, we calculated the reactor size required for a reaction temperature of 25°C. Estimate the reactor volume for a reaction temperature of 50°C using the following assumptions:

$$k_0 = 5.2 \times 10^{13} \ L/gmol \ s$$
$$E_a = 3500 \ cal/gmol$$

8. As a chemical engineer working for MedAid, Inc., you are assigned to size a reactor for producing Agent X, a new chemical designed to combat killer viruses. X is produced by the following second-order liquid-phase reaction:

$$A + Z \rightarrow X$$

where the reaction rate is described by

$$r_{reaction,A} = k_r c_A c_Z$$

The activation energy for the reaction is 31.5 *kcal/gmol*, and k_o = 2.0 x 10^{16} *L/gmol s.* Species A (MW = 102) and species Z (MW = 76) are to enter the reactor at a rate of 22 *gmol/s* and 27 *gmol/s*, respectively. An additional inert (non-reacting) component, species I (MW = 25), also enters the reactor at a rate of 38 *gmol/s*. The desired conversion of A in the reactor is 80%. A single product stream leaves the reactor and has a density (as a function of temperature) which is described by:

$$\rho_{product} \ (g/cm^3) = 1.2 - 0.0012 \ (T\text{-}298)$$

where T is the temperature in units of K.

a. Use a spreadsheet to calculate the necessary reactor volume as a function of temperature for a range of temperatures from 298 K to 375 K. The calculation should be performed for at least 15 temperatures in this range. Your answer should include a table (printed from the spreadsheet) which provides values of k_r, $\rho_{product}$, $\dot{V}_{product}$, $r_{reaction,A}$, and $V_{reactor}$ (in *liters*) as a function of temperature. Note that the molar and mass flow rates are not a function of temperature.

b. Does the required reactor volume increase or decrease with increasing temperature? Why?

c. At what temperature (within the specified range) would you recommend operating the reactor? Why? What is the required reactor volume at this temperature?

d. Is it feasible to operate the reactor at room temperature? Why or why not?

e. What practical or physical considerations not included in the calculations might limit or influence the practical operating temperature of the reactor? (list at least two)

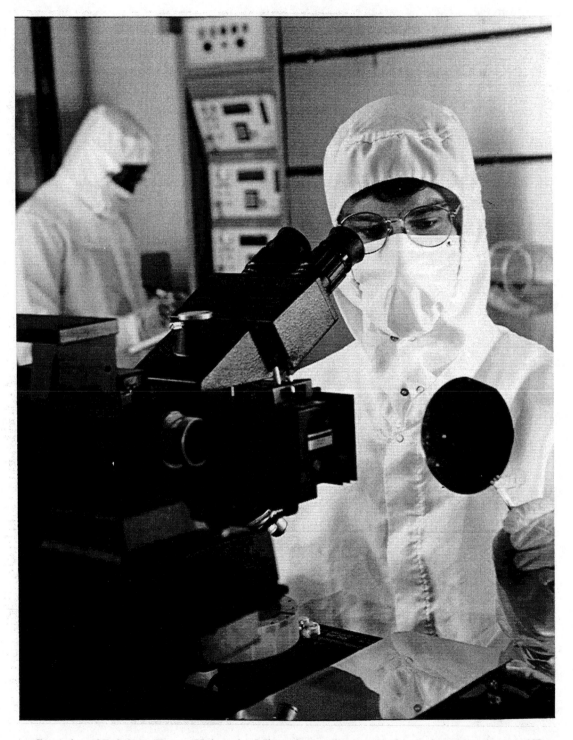

Courtesy of Brigham Young University Microelectronics Processing Laboratory, Provo, UT

CHAPTER 10

HEAT TRANSFER
(COOLING DOWN THE PRODUCT)

Section 10.1   Energy Balances for Steady-State Open Systems

Up to now, we have developed a strategy to neutralize the process acid and have formulated some details of the neutralization system, but we haven't dealt with the temperature of the neutralized acid. Let's suppose that some data from the company indicate an average temperature of $60°C$ $(140°F)$ for the HCl byproduct as it comes from the manufacturing process. The state requires that effluent discharged to a natural river or lake cannot exceed $27°C$ in temperature. These facts lead to several questions:

1. Will the temperature of the mixture still exceed the government's maximum-temperature requirement even after the addition of the NaOH?

2. If the answer to question #1 is "Yes"...

   a. How much heat would still need to be removed to bring the temperature down below the government's maximum-temperature requirement?

   b. What strategy should be used to remove the heat referred to in question 2a?

   c. What other resources are needed to implement the strategy referred to in question 2b?

To answer the important questions posed above, we first have to learn how to describe changes in the energy of a material and the associated changes in temperature.

In your chemistry and/or physics classes, you were probably introduced to the First Law of Thermodynamics for a closed system (a fixed volume or space with no streams entering or leaving the system). The First Law for that system is:

$$\Delta E = Q + W \qquad (10.1)$$

where

E = total energy of the system (units of energy)
Q = heat transferred from the environment to the system through the boundaries of the volume (units of energy) over the interval of time during which E changes
W = work done on the system by the environment (units of energy) over the interval of time during which E changes

The important message of Equation 10.1 is that <u>energy is conserved</u>. While one form of energy may be converted to or from other forms of energy, it will never be created from nothing or disappear to nothing. If work is done on a system, energy is utilized to provide that work, and the energy shows up as increased energy within the system. Similarly, if heat passes across a system boundary into a system, it likewise shows up as increased energy within the system.

A similar balance can be written for *open* systems, that is systems with streams entering and leaving. In this chapter, we restrict our attention to open systems which are also *steady-state systems*, that is systems where the energy of the material does not change with time. With these restrictions, the First Law of Thermodynamics can be written as

$$\underset{\substack{output \\ streams}}{\sum \{\dot{m}\hat{E}\}_{out}} - \underset{\substack{input \\ streams}}{\sum \{\dot{m}\hat{E}\}_{in}} = \dot{Q} + \dot{W} \tag{10.2}$$

where

$\dot{m}$ = mass flow rate of a stream (units of mass per time)

$\hat{E}$ = energy of a stream of flowing material (units of energy per mass of fluid)

$\dot{Q}$ = rate of transfer of energy across the boundaries of a system into that system (units of energy per time)

$\dot{W}$ = rate that work is done on a system (units of energy per time)

Briefly, Equation 10.2 states that, for a steady-state open system, when the energy of the outlet streams is greater than that of the inlet streams, that will be because heat was transferred into the system through the boundaries and/or because work was done on the system by the environment. Let's explore the concept of energy a little further.

Energy (E):

In Chapter 7, we learned about two forms of energy (expressed per unit mass of material):

Kinetic energy: $\frac{1}{2}\alpha v^2$

Potential energy: gz

We now add a third form of energy, the "internal" energy per unit mass (usually given the symbol $\hat{U}$). Just as a flowing stream has kinetic and potential energy, individual molecules also have kinetic energy (from their individual motion) and potential energy (from the attraction and repulsion between molecules). The sum of these molecular energies is expressed as the internal energy of the material, which is a strong function of temperature.

Now, the total energy per mass of material is

$$\hat{E}_{total} = \hat{E}_{internal} + \hat{E}_{kinetic} + \hat{E}_{potential} = \hat{U} + \frac{1}{2}\alpha v^2 + gz \tag{10.3}$$

The units of energy are the same as you have studied in your chemistry and physics courses, and are summarized in Table 10.1.

Table 10.1 Units of Energy

System of Units	Energy Unit	Definition/conversion
cgs	*erg*	$1 \; erg \equiv 1 \; g \; cm^2/s^2$
	calories (*cal*)	$1 \; cal = 4.182 \times 10^7 \; erg$
SI	Joules (*J*)	$1 \; J \equiv 1 \; kg \; m^2/s^2$
American	British thermal unit (*Btu*)	$1 \; Btu = 1055 \; J$

Finally, we now can substitute our definition of $\hat{E}$ (Equation 10.3) into our equation for open systems (Equation 10.2), giving:

$$\sum_{\substack{output \\ streams}} \dot{m}\left[\hat{U}+\frac{1}{2}\alpha v^2 + gz\right]_{out} - \sum_{\substack{input \\ streams}} \dot{m}\left[\hat{U}+\frac{1}{2}\alpha v^2 + gz\right]_{in} = \dot{Q}+\dot{W} \qquad (10.4)$$

At this point, you are probably wondering where all this is headed. In a nutshell, we need some way of calculating the transfer of heat, and thus the temperature, of streams in a chemical process. The energy balance just derived will form the basis for the relationships needed to perform these calculations. Before proceeding further, let's examine more closely a couple of quantities which appear in the steady-state energy balance, $\dot{Q}$ and $\dot{W}$.

Rate of Heat Transfer ($\dot{Q}$):

In Equations 10.2 and 10.4, $\dot{Q}$ represents the rate of heat transfer (energy per time) entering (or leaving) an open system through the system boundaries. The major mechanisms of this heat transfer are *conduction* transfer, *convection* transfer, and *radiation* transfer.

conduction: Heat will transfer through a stationary medium when one location in the medium is at a higher temperature than another location (Figure 10.1).

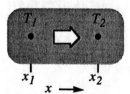

Figure 10.1 Heat transfer by conduction in the "x" direction when the temperature in one region (T_1) is greater than in another (T_2)

One example of such transfer would be inside the metal at the bottom of a cooking pan. Heat transfer by conduction occurs as molecules transfer energy to their neighboring molecules through vibrational collisions. The rate of energy transfer by conduction in a certain direction (let's call it the "x" direction) is described by Fourier's Law of Heat Conduction, which looks a great deal like Ficks Law for mass transfer:

$$\dot{Q}_{cond,x} = -kA\frac{T_2 - T_1}{x_2 - x_1} = -kA\frac{\Delta T}{\Delta x} \qquad (10.5)$$

where

$\dot{Q}_{cond,x}$ = conduction heat-transfer rate (energy transferred per time, e.g. J/s, through area "A") in the "x" direction between locations "1" and "2"

A = cross-sectional area through which conduction occurs

k = the thermal conductivity of the material, with units of energy per time per length per temperature (e.g. $W/m\,K$) (Note: The unit W or $Watts$ is equivalent to the units of energy per time)

T = temperature

The thermal conductivity, k, expresses how easily heat conducts through a material, so a higher value of k leads to more heat transfer, and a lower value leads to less heat transfer. From your experience, you will already recognize that some materials, such as metals, have high thermal conductivities, and other materials, such as those used for insulation, have low values. From representative values of the thermal conductivity given in Table 10.2, it is clear that the best insulator is air. In fact, insulation materials such as fiberglass and double-pane windows derive their insulation value by trapping small pockets of air.

Table 10.2 Approximate Values of k ($W/m\ K$) @ 300 K (from Reference 1)

Air	.026
Wood	.12
Plaster Board	.17
Water	.61
Brick	.72
Glass	1.4
Aluminum	237

The ratio $\Delta T/\Delta x$ in Equation 10.5 represents the rate of change of temperature as we move in the "x" direction. In other words, there will be more conduction if there is a large temperature difference over a small distance. Furthermore, the negative sign in Equation 10.5 indicates that if $\Delta T/\Delta x$ is positive (the temperature is increasing in the x direction), the heat will conduct in the negative-x direction (it will conduct in the direction of decreasing temperature).

convection: If a fluid flows along a surface and is at a temperature different from the temperature of that surface, heat will transfer to or away from that surface (Figure 10.2). The rate of convection heat transfer will increase as the velocity of the fluid increases.

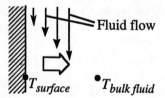

Figure 10.2 Heat transfer by convection from a hot surface to a cooler flowing fluid

Thus, during a cold winter day, our bodies lose heat faster when the wind is blowing than when there is no wind. The mathematical expression used to describe the rate of heat convection is very similar to the law describing mass convection (see Chapter 8), namely:

$$\dot{Q}_{conv} = h\,A\,(T_{surface} - T_{bulk\ fluid}) \qquad (10.6)$$

where

$\dot{Q}_{conv}$ = convection heat-transfer rate (energy transferred per time, e.g. J/s, through area "A") from location "1" to location "2"

h = heat-transfer coefficient, including the effects of the system geometry, the nature of the fluid, and the external forces on the fluid (units of energy per time per area per temperature (e.g. $W/m^2 K$)

A = cross-sectional area through which convection occurs

T = temperature

In your later courses in heat transfer, you will learn how the heat-transfer coefficient is estimated for a given surface geometry and given fluid flow conditions.

radiation: All surfaces radiate heat in the form of electromagnetic waves (Figure 10.3).

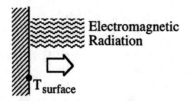

Figure 10.3 Heat transfer by radiation from a surface

Heat transfer by radiation does not require a medium or material for the transfer to take place and can even take place through a vacuum (such as the sun radiating its heat through outer space to the earth). The rate of transfer from a given surface to its surroundings is approximately described by Equation 10.7.

$$\dot{Q}_{rad} = \varepsilon \, \sigma A \, (T_{surface})^4 \tag{10.7}$$

where

$\dot{Q}_{rad}$ = radiation heat-transfer rate (energy transferred per time, e.g. *J/s*, through area "*A*") from location "1" to location "2"

ε = emissivity, which indicates how well the surface emits radiation compared with a "perfect" radiator (unit-less)

σ = Stephan-Boltzman constant (equals 5.669×10^{-8} *W/m²K⁴*)

A = cross-sectional area through which convection occurs

T = temperature

The radiation from a surface depends on the fourth power of the <u>absolute</u> (e.g. units of Kelvin) surface temperature. This fourth-power dependence is the reason that radiation from a surface is not very large at temperatures around room temperature or lower but rapidly becomes important as surface temperatures become significantly higher than room temperature.

It's important to note that, while a surface radiates energy to the environment around it, the environment radiates energy back to that surface, and the net transfer is the difference between those two amounts. When a surface radiates more heat away than it receives from other surfaces, then a net transfer <u>from</u> the surface takes place by radiation, and when a surface receives more heat from other surfaces than it radiates away, a net transfer <u>to</u> the surface takes place by radiation. This is why we feel warmer in the sun than we do in the shade and why the black asphalt (which receives more radiation than it gives off) is so hot on a summer day. The actual calculation of the transfer between surfaces and their environment is too complex for this introductory text and will not be treated here.

In Equation 10.2, $\dot{Q}$ is the <u>net</u> transfer of heat <u>into</u> the system across the boundaries (there may be some transfer in and some transfer out). So, if the heat transfer across the boundaries by conduction, convection, and radiation adds up to a net input, $\dot{Q}$ will be positive. If, however, the total heat transfer from these mechanisms is a net output from the system, $\dot{Q}$ will be negative.

Note that both the thermal conductivity (k) (Equation 10.5) and the heat-transfer coefficient (h) (Equation 10.6) are multiplied by the difference of two temperatures. Therefore, the temperature designation in the denominator of the units of k and h is a temperature *difference*, and the conversion of a <u>temperature difference</u> to a different scale is different from the conversion of a <u>temperature</u>, as illustrated in Figure 10.4. For example, the value for that difference is the same whether expressed in the *Centigrade* or the *Kelvin* scale (remember, temperature in *Kelvin* equals temperature in degrees *Centigrade* plus 273.15, see Figure 10.4). But the temperature difference in *Fahrenheit* is equal to 1.8 times the temperature difference in *Centigrade* as shown in Figure 10.4. In other words, the conversion factors for converting differences are

$$\frac{1°C\ (difference)}{1\ K\ (difference)} \quad \text{and} \quad \frac{1.8°F\ (difference)}{1°C\ (difference)}$$

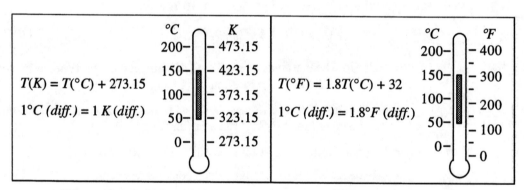

Figure 10.4 Relationship between temperatures and temperature changes

The following example illustrates the conversion of temperature differences:

Example 10.1

A typical value for the thermal conductivity of steel is   53 *W/m K*. What is the corresponding value in units of *Btu/hr ft °F?*

<u>Solution:</u>

$$k = 53\frac{W}{m\ K}\left(\frac{9.478 \times 10^{-4}\ Btu/s}{W}\right)\left(\frac{0.3048m}{ft}\right)\left(\frac{3600s}{hr}\right)\left(\frac{1\ K\ (difference)}{1°C\ (difference)}\right)\left(\frac{1°C\ (difference)}{1.8°F\ (difference)}\right)$$

$$= 31\frac{Btu}{hr\ ft°F}$$

Rate of Work ($\dot{W}$):

When external forces do work on a fluid, the energy of that fluid increases. For example, when a pump does work on a fluid, that work increases fluid velocities, potential energy, and/or fluid temperature. Fluids can also do work on their environment and thereby lose energy.

With open systems, we usually consider two kinds of work, *shaft work* and *flow work*. These are defined as follows:

rate of shaft work ($\dot{W}_s$): When the flowing fluid in a system contacts moving parts, work is performed. This is the same shaft work introduced in Chapter 7 and used in the mechanical energy equation, except that we now express it in units of energy per time and use a dotted capital "$\dot{W}$" as compared with the lower-case "w" used in Chapter 7 to represent energy per mass of fluid. As in Chapter 7, shaft work for our energy balance is defined as work done <u>on</u> the system and is positive when the net work is indeed on the system (such as in a pump, where the moving parts are driven by external forces and thereby "push" the fluid). Similarly, the shaft work has a negative value when the net work is done <u>by</u> the fluid (such as in a turbine, where the fluid <u>causes</u> the parts to move).

rate of flow work ($\dot{W}_{PV}$): This work results from the displacement of fluid during flow and is similar to the pressure-volume work associated with the compression or expansion of a closed system. However, in the case of an open system (with inlet and outlet streams), the flow of fluid into and out of a system represents a continual performance of work as upstream fluid "pushes" fluid into the system entrance and the fluid in the system "pushes" downstream fluid out of the system exit. This work is calculated from Equation 10.8:

$$\dot{W}_{PV} = \underset{\substack{input \\ streams}}{\sum}\left(P\dot{V}\right)_{in} - \underset{\substack{output \\ streams}}{\sum}\left(P\dot{V}\right)_{out} = \underset{\substack{input \\ streams}}{\sum}\left(\dot{m}P\hat{V}\right)_{in} - \underset{\substack{output \\ streams}}{\sum}\left(\dot{m}P\hat{V}\right)_{out} \qquad (10.8)$$

where $\hat{V}$ is the volume per unit mass of the stream (which, incidentally, also equals $1/\rho$).

Summarizing this description of work, we will say that

$$\dot{W} = \dot{W}_s + \dot{W}_{PV} \qquad (10.9)$$

Inserting the information from Equation 10.9 into the open-system energy balance (Equation 10.4) gives us

$$\underset{\substack{output \\ streams}}{\sum}\left\{\dot{m}\left[\hat{U} + \tfrac{1}{2}\alpha v^2 + gz\right]\right\}_{out} - \underset{\substack{input \\ streams}}{\sum}\left\{\dot{m}\left[\hat{U} + \tfrac{1}{2}\alpha v^2 + gz\right]\right\}_{in} = \dot{Q} + \dot{W}_s + \dot{W}_{PV} \qquad (10.10)$$

Further substitution of Equation 10.8 into Equation 10.10 and rearrangement gives

$$\underset{\substack{output \\ streams}}{\sum}\left\{\dot{m}\left[\hat{U} + P\hat{V} + \tfrac{1}{2}\alpha v^2 + gz\right]\right\}_{out} - \underset{\substack{input \\ streams}}{\sum}\left\{\dot{m}\left[\hat{U} + P\hat{V} + \tfrac{1}{2}\alpha v^2 + gz\right]\right\}_{in} = \dot{Q} + \dot{W}_s \qquad (10.11)$$

It turns out that Equation 10.11 is a more complete expression for the energy balance, of which the mechanical energy equation introduced in Chapter 7 is a subset. Note that the term $P\hat{V}$ in Equation 10.11 is the same as the term P/ρ in Equation 7.8. Also, it turns out that w_f can be derived from $\hat{U}$ and $\dot{Q}$.

The internal energy is often combined with the flow work into a property called the "enthalpy," represented by the symbol $\hat{H}$ and defined as

$$\hat{H} = \hat{U} + P\hat{V} \tag{10.12}$$

Substituting Equation 10.12 into 10.11 results in the most common form of the *steady-state open-system energy balance*:

$$\sum_{\substack{output \\ streams}} \left\{ \dot{m}\left[\hat{H} + \tfrac{1}{2}\alpha v^2 + gz\right] \right\}_{out} - \sum_{\substack{input \\ streams}} \left\{ \dot{m}\left[\hat{H} + \tfrac{1}{2}\alpha v^2 + gz\right] \right\}_{in} = \dot{Q} + \dot{W}_s \tag{10.13}$$

Section 10.2   Some Applications of the Steady-State Energy Balance

Equation 10.13 represents a general form of the steady-state energy balance which is applicable to a broad range of problems and is applied extensively in upper-level chemical engineering courses. In this section, we will consider some simple applications, and we will begin by deriving a simplified form of Equation 10.13. First of all, for problems where we are trying to determine $\hat{H}$ or $\dot{Q}$, the magnitudes of the kinetic energy and potential energy terms are typically small compared with the other terms of Equation 10.13 and therefore can be neglected. In addition, the applications we will consider do not include the performance of shaft work, so our simplified balance will not include $\dot{W}_s$. Hence, a <u>steady-state energy balance with negligible change in kinetic and potential energies and with no shaft work</u> is

$$\sum_{\substack{output \\ streams}} \left(\dot{m}\hat{H}\right)_{out} - \sum_{\substack{input \\ streams}} \left(\dot{m}\hat{H}\right)_{in} = \dot{Q} \tag{10.14}$$

Equation 10.14 can be used in processes where phenomena such as heating/cooling, phase changes, and/or chemical reactions occur simultaneously. However, for the purpose of this introductory book, we will limit our treatment to processes where only one of these phenomena is occurring. We will therefore simplify Equation 10.14 to develop a specific equation for each of the following: 1) sensible heating or cooling, 2) phase changes, and 3) chemical reactions.

Scenario #1: Sensible heating or cooling

When a material is warmed or cooled without a phase change, we call this process *sensible heating or cooling*. Figure 10.4 depicts sensible heating; note that $\dot{Q}$ is positive when heat is added to the process. How would Figure 10.4 be changed to represent sensible cooling?

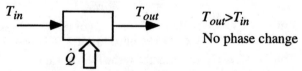

Figure 10.5 Sensible heating

To use Equation 10.14, it is necessary to calculate the enthalpy of the input and output streams. For sensible heating/cooling, the specific enthalpy, $\hat{H}$, can be approximated as

$$specific\ enthalpy = \hat{H} \approx \overline{C}_p(T - T_{ref})$$ (10.15)

where $\overline{C}_p$ is the *heat capacity* of the fluid averaged over the temperature range from a reference temperature, T_{ref}, to the temperature T. The heat capacity expresses the amount of energy needed to raise the temperature of a unit mass (e.g. 1 *gram*) of the fluid by a defined amount (e.g. 1.0°C) at constant pressure. The units of $\overline{C}_p$ are energy per mass per temperature change, such as *cal/g°C* or *BTU/lb_m°F*. Different substances have different heat capacities. For example, the heat capacity of water at 25 °C equals 1 *cal/g°C*. Materials with large heat capacities (e.g. most solids and liquids) require large amounts of heat to raise their temperature. In contrast, much less heat is required to produce the same increase in temperature in materials with small heat capacities (e.g. most gases). In the units of heat capacity, the temperature unit in the denominator is a temperature change and must be handled in the same way as discussed for the units of k and h (see Figure 10.4).

The reference temperature is typically chosen to be 25°C and is the temperature at which the value of the enthalpy is defined as zero. Therefore, *for all problems in this book, we will assume that T_{ref} equals 25 °C for all substances.*

Substituting Equation 10.15 into Equation 10.14 gives us our <u>steady-state energy balance for sensible heating or cooling</u>:

$$\boxed{\sum_{\substack{output \\ streams}} \dot{m}\overline{C}_p(T - T_{ref}) - \sum_{\substack{input \\ streams}} \dot{m}\overline{C}_p(T - T_{ref}) = \dot{Q}}$$ (10.16)

Equation 10.16 is typically used to determine the amount of heat that must be transferred to achieve a specified amount of sensible heating/cooling or, alternately, for determining one or more temperatures which would result from a given heat addition/removal. It is often used in conjunction with material balances, since material balances may be needed to determine unknown flow rates in a heat-transfer problem, as illustrated in the following example:

Example 10.2

Motor oil is being blended in a steady-state process where the feed streams to the process are two oil stocks and an additive. The properties of the feed streams are:

	Mass flow rate (kg/min)	Heat Capacity (kJ/kg K)	Temperature (°C)
Oil #1	18.3	2.11	105
Oil #2	13.9	2.32	112
Additive	1.4	1.87	93

The heat capacity of the resulting product stream equals 2.19 *kJ/kg K*, and the reference temperature for all streams is 25 °C.

a. If no heat is added or removed from the process, what is the temperature of the product stream?

b. If the temperature of the product stream must be 78 °C, how much heat must be added or removed from the process?

Solution: The drawing for this problem is

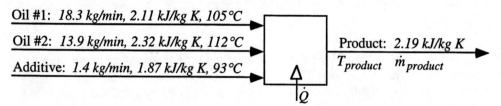

Oil #1: *18.3 kg/min, 2.11 kJ/kg K, 105°C*
Oil #2: *13.9 kg/min, 2.32 kJ/kg K, 112°C*
Additive: *1.4 kg/min, 1.87 kJ/kg K, 93°C*

Product: *2.19 kJ/kg K*
$T_{product}$ $\dot{m}_{product}$

$\dot{Q}$

We begin with a material balance such as in Equation 5.3, which tells us that

$$\dot{m}_{product} = \dot{m}_{out} = \sum \dot{m}_{in} = 18.3 + 13.9 + 1.4 = 33.6 \frac{kg}{min}$$

a. In the case where no heat is added or removed, the term $\dot{Q}$ in Equation 10.16 equals 0 (we call this kind of process adiabatic), so substituting the other values gives

(33.6 *kg/min*)(2.19 *kJ/kg K*)($T_{product}$ - 25°C) - (18.3 *kg/min*)(2.11 *kJ/kg K*)(105 - 25°C)
- (13.9*kg/min*)(2.32 *kJ/kg K*)(112 - 25°C) - (1.4 *kg/min*)(1.87 *kJ/kg K*)(93 - 25°C) = 0

which gives

(33.6 *kg/min*)(2.19 *kJ/kg K*)($T_{product}$ - 25°C) - 6073 kJ/min = 0; $T_{product}$ = 108°C

Notice: Because the *Kelvin* unit in the denominator of the heat capacity is a temperature <u>difference</u>, its value is the same as a difference in degrees Centigrade. Hence it cancelled the *Centigrade* temperature unit when multiplied by the temperature.

b. In the case where the product temperature must be 78°C and heat may be added or removed, Equation 10.16 becomes

(33.6 *kg/min*)(2.19 *kJ/kg K*)(78 - 25°C) - (18.3 *kg/min*)(2.11 *kJ/kg K*)(105 - 25°C)
- (13.9*kg/min*)(2.32 *kJ/kg K*)(112 - 25°C) - (1.4 *kg/min*)(1.87 *kJ/kg K*)(93 - 25°C) = $\dot{Q}$

which gives $\dot{Q}$ = -2173 *kJ/min*

The negative value for $\dot{Q}$ indicates that heat must be removed, so we conclude that 2173 *kJ/min* must be removed from the process.

Scenario #2: Phase changes

Chemical processes often include the change of a material between the three phases: solid, liquid, and vapor. For example, water can be in the form of solid (ice), liquid, or vapor (steam). Such processes involve the addition or removal of heat (such as adding heat to melt ice).

The three combinations of phases between which a change can occur are as follows:
1. liquid to vapor (vaporization) or vapor to liquid (condensation)
2. solid to liquid (melting) or liquid to solid (freezing)
3. solid to vapor (sublimation) or vapor to solid (solid condensation)

For a single-component pure substance, such a change in phase takes place at constant temperature and can be depicted as shown in Figure 10.6.

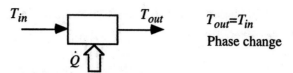

Figure 10.6 Phase change

Each phase change occurs at a specific temperature (boiling point, melting point, sublimation point) and will not take place unless the material is at that temperature. For example, the addition of heat to a liquid will increase its temperature (sensible heating) until it reaches its boiling point. Once the boiling point has been reached, the temperature of the liquid remains constant, and the addition of any more heat causes the liquid to vaporize (boil). In other words, any additional energy is used to vaporize the liquid (i.e. change the phase from liquid to vapor) rather than to raise its temperature. A pure liquid at its boiling is called a *saturated liquid*. Similarly, a vapor cooled to its boiling point becomes a *saturated vapor*, since the removal of additional heat will cause condensation. Note that some of the phase-change temperatures described above, e.g. the boiling point, vary significantly with pressure.

Each change in phase also is associated with a change in the specific enthalpy ($\hat{H}$) of the material. The enthalpy change has various names, depending upon the phases involved and the historical development of such names; some common names are included in Table 10.3. In each case, the $\Delta\hat{H}$ term usually has units of energy per mass of material undergoing the phase change (it may also be expressed per mole of material). The enthalpy changes and temperatures discussed in this paragraph and the preceding paragraph are summarized in Table 10.3.

Table 10.3 Common Terminology Associated with Changes in Phase

Process	Common names for temperature of phase change	$\Delta\hat{H}_{phase\ change}$
Vaporization (boiling)	boiling point	$\Delta\hat{H}_{vap}$
Condensation		$(-\Delta\hat{H}_{vap})$
Melting	melting point or freezing point	$\Delta\hat{H}_{fusion}$ or $\Delta\hat{H}_m$
Freezing		$(-\Delta\hat{H}_{fusion}$ or $-\Delta\hat{H}_m)$
Sublimation	sublimation point	$\Delta\hat{H}_s$
Solid condensation		$(-\Delta\hat{H}_s)$

In this book, we will only consider processes involving phase changes with only one inlet stream and one outlet stream (as shown in Figure 10.6) and where the phase change occurs isothermally at the temperature for which we have a value for $\Delta\hat{H}_{phase\ change}$. In such cases,

$$\hat{H}_{out} - \hat{H}_{in} = \Delta\hat{H}_{phase\ change} \quad\quad\quad (10.17)$$

where $\Delta\hat{H}_{phase\ change}$ is in units of energy per mass, the same as for $\hat{H}$. Substituting Equation 10.17 into Equation 10.14 gives us our <u>steady-state energy balance for phase change</u>:

$$\dot{m}_{phase\ change}\ \Delta\hat{H}_{phase\ change} = \dot{Q} \quad\quad\quad (10.18)$$

where $\dot{m}_{phase\ change}$ is the mass flow rate of material undergoing the phase change (which might not equal the total mass flow rate into the process). Note also that we must be careful how we relate $\Delta\hat{H}_{phase\ change}$ to available data and how we interpret $\dot{Q}$. For example, if we are vaporizing water where the inlet stream is water and the outlet stream is steam at the same temperature, $\Delta\hat{H}_{phase\ change} = \Delta\hat{H}_{vap}$, and $\dot{Q}$ has a positive value, since we must add heat to the system. On the other hand, if we are condensing steam to form water, $\Delta\hat{H}_{phase\ change} = -\Delta\hat{H}_{vap}$, and heat must be removed ($\dot{Q}$ has a negative value). Also remember that Equation 10.18 applies to systems with only a single inlet and single outlet stream.

Typically, we use Equation 10.18 to determine the amount of heat transfer ($\dot{Q}$) associated with a phase change for a certain flow rate of material, or we use it to determine the amount of material (e.g. $\dot{m}$) which must undergo a phase change to match a certain rate of heat transfer. For example, we might want to calculate the amount of heat required to generate a desired amount of steam, as shown in the following example:

Example 10.3

A steady-state boiler produces steam from a waste-water stream. The water enters the boiler as saturated water at 5.7 *atm* and 430 *K* (the boiling point of water at 5.7 *atm*), and the steam exits the boiler as saturated steam under that same temperature and pressure. The properties of importance are:

Mass flow rate:	8150 *kg/hr*
Heat of vaporization at 5.7 *atm* and 430 *K*:	2091 *kJ/kg*

How much heat must be added to the process? (Notice that the boiling point and heat of vaporization are dependent on pressure, so it was necessary to know their values for the operating pressure of the process.)

<u>Solution</u>: The diagram for this process is

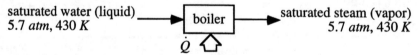

Equation 10.18 can be applied directly, yielding

$$(8150\ kg/hr)(2091\ kJ/kg) = \dot{Q}$$

or

$$\dot{Q} = 17 \times 10^6\ kJ/hr$$

Notice that heat is being <u>added</u> to the process, so $\dot{Q}$ is a positive value.

Scenario #3: Chemical reactions

Most chemical reactions produce an enthalpy change, designated as $\Delta\tilde{H}_{reaction,A}$, expressed here in units of energy per *moles* of species *A* reacted (it could also be expressed per *mass* of species *A* reacted, but the equation we derive would be slightly different). Positive values of $\Delta\tilde{H}_{reaction,A}$ indicate that the reaction absorbs heat - i.e. the reaction is *endothermic*. Similarly, negative values of $\Delta \tilde{H}_{reaction,A}$ indicate that the reaction gives off heat - i.e. the reaction is *exothermic*. In this book, we will only consider processes where only one reaction occurs and where the reaction occurs isothermally at the temperature for which we have a value for $\Delta \tilde{H}_{reaction,A}$.

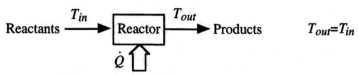

Figure 10.7 Chemical reaction

In such cases, the left-hand side of Equation 10.14 becomes

$$\sum_{\substack{output \\ streams}} \left(\dot{m}\hat{H} \right)_{out} - \sum_{\substack{input \\ streams}} \left(\dot{m}\hat{H} \right)_{in} = \left(\frac{moles\ of\ A\ reacted}{time} \right) \Delta \tilde{H}_{reaction,A} = r_{consumption,A} \Delta \tilde{H}_{reaction,A}$$

where $r_{consumption,A}$ is determined from material balances as discussed in Chapter 5. Setting this equal to the right-hand side of Equation 10.14, we obtain our <u>steady-state energy balance for chemical reaction</u>:

$$\longrightarrow \boxed{r_{consumption,A} \Delta \tilde{H}_{reaction,A} = \dot{Q}} \longleftarrow \qquad\qquad (10.20)$$

$\dot{Q}$ in Equation 10.20 represents the rate at which heat is produced or consumed by the chemical reaction. One use of Equation 10.20 would be to determine the amount of heat which must be removed or added to maintain the reactor at a constant temperature.

Example 10.4

Toluene at 1200 °F is fed to a reactor at 373 *lbmol/hr* where 75% of it reacts with hydrogen to form benzene by the following reaction:

$$Toluene + H_2 \rightarrow Benzene + CH_4$$

The H_2 enters the reactor as a separate feed stream (also at 1200 °F) at three times the rate required to react with all of the toluene. The reactor is maintained at 1200 °F and 500 *psia* , and the heat of reaction is -21,540 *Btu/lbmol* of toluene reacted.

a) In which direction (addition or removal) would heat need to be transferred relative to the reactor to keep the temperature constant in the reactor?

b) How much heat (expressed as *Btu/hr*) would need to be added or removed?

Hint: You may not need all of the above information to solve this problem.

<u>Solution</u>: The diagram is

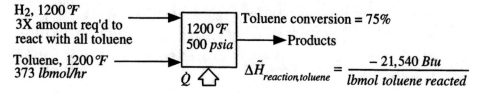

a) Since the value of $\Delta \tilde{H}_{reaction, toluene}$ is negative, the reaction is exothermic. Thus, we will need to <u>remove</u> heat to prevent the temperature from increasing.

b) From the relationship between conversion and input given in Table 5.2

$$r_{consumption, toluene} = X \sum_{\substack{input \\ streams}} \dot{n}_{toluene, in} = (0.75)\left(373\frac{lbmol}{hr}\right) = 280\frac{lbmol}{hr}$$

Equation 10.20 then becomes

$$\left(280\frac{lbmol}{hr}\right)\left(-21,540\frac{Btu}{lbmol}\right) = -6.03x10^6 \frac{Btu}{hr} = \dot{Q}$$

Now, let's apply what we've learned to the acid-neutralization problem. We will begin by reviewing the combining of the HCl and NaOH streams originally presented in Figure 9.6. The relationships are shown in Figure 10.8, where concentrations have been replaced by temperatures and the reactor has been replaced by a mixer.

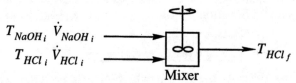

Figure 10.8 The flow rates and temperatures of the HCl, NaOH, and final streams

In this problem, we know that the concentrations of the acid and base are relatively dilute, so we are justified in neglecting the heat effects from the reaction itself. Thus, we will treat this as a mixing of two streams of different temperature, which produces sensible heating of one stream and sensible cooling of the other. Further, we will assume that conduction through the walls and convection and radiation from the walls of the pipes and mixer will not be important. Thus, the $\dot{Q}$ term in Equation 10.16 will be zero. As before, we will assume that kinetic and potential energy considerations will be negligible when compared with the other energy terms (again, this is usually true for liquid streams under typical conditions). Furthermore, there are no mechanisms in the system for major work to be done by the fluid. As a result, Equation 10.16 is directly applicable to this problem and becomes

$$\left\{\dot{V}\rho C_p(T - T_{ref})\right\}_f - \left\{\dot{V}\rho C_p(T - T_{ref})\right\}_{HCl} - \left\{\dot{V}\rho C_p(T - T_{ref})\right\}_{NaOH} = 0 \qquad (10.21)$$

Finally, we can assume that the three streams (NaOH stream, HCl stream, and final stream) are all very dilute, have the same reference temperature, and have approximately the same heat capacity and density as water. Also, from Equation 5.3 (with appropriate substitutions and constant density),

$$\dot{V}_f = \dot{V}_{HCl_i} + \dot{V}_{NaOH_i} \qquad (10.22)$$

You should verify that the application of these assumptions and insertion of Equation 10.22 into Equation 10.21, followed by some rearrangement, gives:

$$T_f = \frac{\dot{V}_{HCl_i} T_{HCl_i} + \dot{V}_{NaOH_i} T_{NaOH_i}}{\dot{V}_{HCl_i} + \dot{V}_{NaOH_i}} \qquad (10.23)$$

We can now answer question #1 posed at the beginning of this chapter, namely whether the temperature of the HCl-NaOH mixture (the reactor outlet) will still exceed the state maximum temperature requirement.

The NaOH solution that we will mix with the acid solution will be stored in a tank, and its temperature will vary as the temperature in the surrounding air varies. Since we want always to be in compliance with the governmental regulation, we should perform all design calculations for the worst case situation. In this case, the worst case will be when the stored NaOH solution is at its highest temperature (say during the warmest part of the summer), which we will estimate as being $32\,°C$ (i.e. $T_{NaOH_i} = 32\,°C$).

Also, from the spreadsheet calculation in Chapter 6, we estimated the average flow rate of NaOH ($\dot{V}_{NaOH_i}$) to be 6500 L/hr.

Substituting these values into Equation 10.23 gives

$$T_f = \frac{(11600\,L/hr)(60°C) + (6500\,L/hr)(32°C)}{11600\,L/hr + 6500\,L/hr} = 49.9°C$$

Since this temperature is still significantly higher than the governmental specification of $27\,°C$ for maximum temperature, it is clear that some additional measure will need to be taken to cool the final mixture to the required temperature.

Summarizing the Procedure for Using the Energy Balance

We apply energy balance equations in the same manner that we learned to apply other general balance equations, such as the material balance equations. The steps have been illustrated in the example problems, but we now present a formal list of steps similar to the one given in Chapter 5:

Table 10.4 Steps for Analyzing Energy Balance Problems

1. Draw a diagram if one is not already available.
2. Write all *known* quantities (flow rates, densities, temperatures, etc.) in the appropriate locations on the diagram. If symbols are used to designate known quantities, include those symbols on the diagram.
3. Identify and assign symbols to important *unknown* quantities and write them in the appropriate locations on the diagram.
4. Determine whether the problem involves sensible heating/cooling, phase change, or chemical reaction and write the appropriate simplified energy balance equation (Equation 10.16, 10.18, or 10.20). Along with the balance equation, write down the given information associated with that equation, such as average heat capacities, enthalpy changes for a phase change, or enthalpy changes of reaction.
5. Construct appropriate material balance equations to aid in determining unknown flow rates or other material-related information. Continue to seek such equations until the total number of equations equals the number of unknowns. In constructing the material balances, use the principles outlined in Chapter 5.
6. Solve the equations to determine the desired unknown quantities.

Section 10.3   Heat Exchange Devices

There are numerous methods for increasing or decreasing the temperature of a fluid. The heating of fluids to extremely-high temperatures is typically accomplished using a fuel-fired furnace through which the fluid is directed. Also, the cooling of fluids to low temperatures usually requires the use of chillers, which operate on the same principles as your household freezer. Both of these methods require significant energy and expense to operate. When more moderate heating or cooling of a stream is required, a cost-effective method is to exchange heat with another stream which is hotter or cooler than the target stream. The other stream can be one which is already part of the process (*process stream*) or can be a stream which is brought in just for heating or cooling (called a *utility stream*), e.g. steam or cooling water.

A *heat exchanger* is a device designed to exchange heat between two flowing streams. In these devices, the streams flow through the exchanger separated by a conducting wall through which the heat can pass from the hot stream to the cold stream. This principle is illustrated in Figure 10.9 in schematic form.

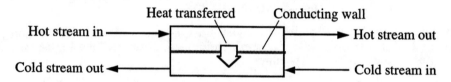

Figure 10.9 Schematic of the principle of heat-exchanger operation

The simplest type of heat exchanger is the concentric-tube heat exchanger (Figure 10.10), which consists of one pipe surrounding another. The figure shows the hot fluid carried by the central pipe and the cold fluid flowing through the annular space outside the inner pipe, but the location of the hot and cold streams could also be the other way around. The exploded view highlights the fact that heat transfer occurs by convection from the hot fluid to the pipe wall, then by conduction through that pipe wall, and then again by convection from the pipe wall into the cold fluid. A variation of this kind of heat exchanger is the shell-and-tube heat exchanger, which has a much larger outer pipe (shell) and a bundle of many tubes passing through that shell. Another variation is the parallel-plate exchanger, in which the cold and hot fluids flow through alternate spaces between plates.

One variable affecting the performance of heat exchangers is the direction of fluid flow. For example, Figure 10.11 shows both fluids entering the exchanger on the same end and flowing in the same direction along the exchanger. This arrangement is called *co-current* or *parallel* flow. In this arrangement, the temperature difference between the two fluids is a maximum at the point where the two fluids enter the exchanger, but that temperature difference decreases significantly as the fluids move through the exchanger (Figure 10.11). The plumbing could be reversed for one of the fluids so that the fluids flowed in opposite directions, called *countercurrent* flow. In this case, the temperature difference between the two fluids does not vary as dramatically as in the co-current exchanger. Furthermore, in countercurrent flow, the outlet of the "cold" fluid could be hotter than the outlet of the "hot" fluid.

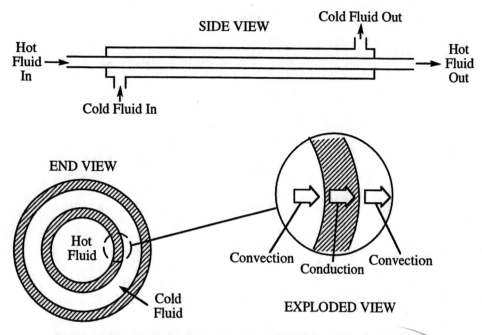

Figure 10.10 An illustration of a concentric-cylinder heat exchanger

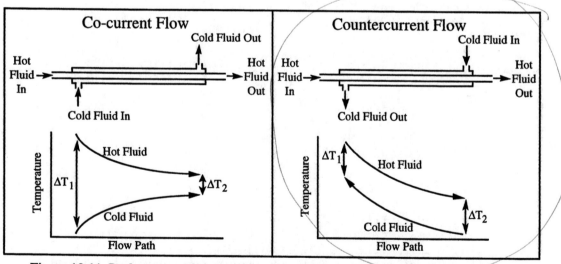

Figure 10.11 Performance of co-current and countercurrent concentric-tube exchangers

In practice, it is often necessary to transfer more heat than is possible with a simple concentric-tube heat exchanger like the one illustrated in Figure 10.10. A shell-and-tube heat exchanger (described previously) is the most common alternative. Figure 10.12 illustrates a shell-and-tube exchanger which is configured in countercurrent mode, but you should notice that the fluid on the shell side actually flows across the outside of the tubes (we call this "crossflow" because of the presence of baffles on the shell side).

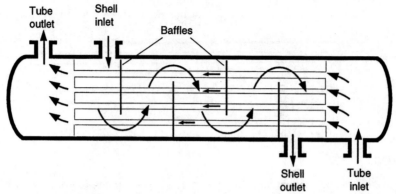

Figure 10.12 Shell-and-tube heat exchanger: one tube pass, one shell pass

The decision of whether to operate a heat exchanger in co-current or countercurrent mode is generally based on which mode yields the smallest heat exchanger to transfer a given amount of heat. Practical issues such as piping layout may also factor into the decision. Most heat exchangers utilize countercurrent flow, and for this introductory treatment, you should assume that a heat exchanger is countercurrent unless otherwise specified.

For the exchanger illustrated in Figure 10.12, the fluid flowing inside the tubes passes from one end of the exchanger to the other end only once, and the fluid on the shell side also passes from one end of the exchanger to other end only once, so we call this a single-tube-pass, single-shell-pass exchanger. It is possible to configure a heat exchanger with more than one tube pass and more than one shell pass, as illustrated in Figure 10.13.

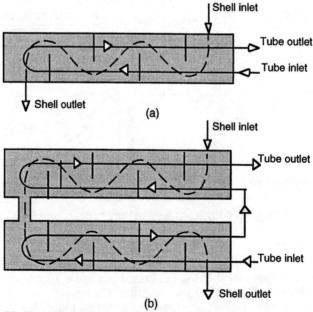

Figure 10.13 Shell-and-tube heat exchangers: a) one shell pass and two tube passes
b) two shell passes and four tube passes

Now that we know what a heat exchanger is, the next step is to learn how to analyze and design such devices. The problem usually begins with a *target stream*, which is a process stream that must be heated or cooled. We typically know the flow rate, inlet temperature, and desired outlet temperature of that target stream. With this information, the strategy is to use the <u>energy balance equations</u> developed in the previous section to determine 1) the amount of heat that must be transferred in the heat exchanger and 2) the flow rate and temperatures of the other stream. We can then determine the required <u>heat-transfer area</u>, which is the most common way of expressing the size of the heat exchanger for preliminary designs.

Applying the Energy Balance Equations to Heat Exchangers

One way of mathematically describing heat exchanger performance is by treating the exchanger as two separate processes, one involving the hot stream and the other involving the cold stream, as illustrated in Figure 10.14.

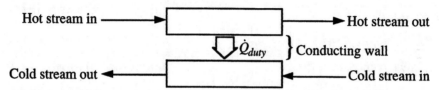

Hot stream in ⟶   ⟶ Hot stream out

$\dot{Q}_{duty}$ } Conducting wall

Cold stream out ⟵   ⟵ Cold stream in

Figure 10.14 Schematic of a heat-exchanger envisioned as two separate processes

As depicted in the figure, heat transfers from the hot fluid at a rate designated as $\dot{Q}_{duty}$ and transfers to the cold fluid at the same rate (thus, we speak of the "duty" of the exchanger). One can construct heat balances for each of the streams as suggested by Equations 10.16, 10.18, and 10.20. The actual form of the balance will depend on whether the stream is undergoing sensible heating/cooling, phase change, or chemical reaction. Again, we will assume that we want this exchanger to operate in <u>steady state</u> (no changes with time), that velocity and elevation changes are unimportant, and that no shaft work is done. The energy balance for each side of the heat exchanger would be formulated as follows for the given situation:

<u>Sensible heating and cooling</u>: Assuming that $\overline{C}_p$ is approximately the same for the stream outlet as for the stream inlet, Equation 10.16 becomes

$$\left[\dot{m}\overline{C}_p (T_{out} - T_{in}) \right]_{hot} = -\dot{Q}_{duty} \tag{10.24a}$$

$$\left[\dot{m}\overline{C}_p (T_{out} - T_{in}) \right]_{cold} = \dot{Q}_{duty} \tag{10.24b}$$

<u>Phase change</u>: Equation 10.18 becomes

$$\left[\dot{m}\Delta\hat{H}_{phase\ change} \right]_{hot} = -\dot{Q}_{duty} \tag{10.24c}$$

$$\left[\dot{m}\Delta\hat{H}_{phase\ change} \right]_{cold} = \dot{Q}_{duty} \tag{10.24d}$$

<u>Chemical reaction</u>: Equation 10.20 becomes

$$\left[r_{consumption,A}\Delta\tilde{H}_{reaction,A} \right]_{hot} = -\dot{Q}_{duty} \tag{10.24e}$$

$$\left[r_{consumption,A}\Delta\tilde{H}_{reaction,A} \right]_{cold} = \dot{Q}_{duty} \tag{10.24f}$$

Note that $\dot{Q}_{duty}$ is, by definition, a positive number, so that the $\dot{Q}$ term in the original energy balance equations became $-\dot{Q}_{duty}$ for the hot stream, because heat is being <u>removed from</u> the hot stream.

For a heat exchanger of interest, the energy balances for the hot stream and cold stream (as per Equation 10.24) can be solved together to determine the unknown temperature, flow rate, or amount of heat transferred. The method relies on the fact that the energy lost by the hot stream as it travels through the exchanger equals the energy gained by the cold stream, as illustrated in Example 10.4 (this assumes that the amount of heat lost to the surroundings is negligible, which is typically the case).

The following example illustrates the use of the energy balance equations to determine the needed flow rate of steam to produce a needed duty.

Example 10.5

A heavy oil stream must be heated to a higher temperature, so the decision is made to use a heat exchanger with saturated steam being condensed to saturated water as the heating source on the other side of the exchanger. The characteristics of the oil are:

Oil mass flow rate:	960 lb_m/min
Oil mean heat capacity:	0.74 $Btu/lb_m\,°F$
Oil inlet temperature:	35°F
Desired oil outlet temperature:	110°F

The saturated steam has the following properties:

Steam temperature:	280°F
Heat of vaporization (@280°F):	925 Btu/lb_m

What steam flow rate is needed for this exchanger?

<u>Solution:</u>

Saturated steam, 280°F, $\dot{m}_{steam}$ ———→ [exchanger] ———→ Saturated water, 280°F, $\dot{m}_{steam}$

Oil, 110°F, 960 lb_m/min ←——— [exchanger] ←——— Oil, 35°F, 960 lb_m/min

For this problem, the oil is the cold stream, and the steam/water is the hot stream. For the oil side, Equation 10.24b gives

$$\dot{Q}_{duty} = \left[\dot{m}\overline{C}_p(T_{out} - T_{in})\right]_{oil}$$

$$= \left(960\frac{lb_m}{min}\right)\left(0.74\frac{Btu}{lb_m\,°F}\right)(110 - 35°\,F) = 53,280\frac{Btu}{min}$$

For the steam/water side, as indicated in Table 10.3, for condensation

$$\Delta\hat{H}_{phase\,change} = -\Delta\hat{H}_{vaporization}$$

so Equation 10.24c gives

$$\dot{m}_{steam} = \frac{-\dot{Q}_{duty}}{-\Delta\hat{H}_{vap}} = \frac{-53,280\,Btu/min}{-925\,Btu/lb_m} = 57.6\frac{lb_m}{min}$$

Determining the Size (Heat-Transfer Area) of the Exchanger

An important characteristic of a heat exchanger is the amount of area available through which the heat can transfer from the hot to the cold side. For example, in a shell-and-tube exchanger, that area is the total surface area of the tubes. The area that a particular heat exchanger must have in order to produce the desired heat transfer depends on how quickly each step in the transfer (convection in the hot fluid, conduction through the wall, and convection in the cold fluid) takes place. Furthermore, to estimate these transfer rates requires an estimation of the heat-transfer coefficients in the hot and cold fluids. That estimation will not be treated in this introduction, as it requires more complete understanding of the effects of surface geometries and flow conditions and will be the subject of more advanced classes. However, an approximate method of estimating the required heat-transfer area will now be developed.

As illustrated in Figure 10.10, the transfer of heat in a heat exchanger is composed of at least three steps in series (one-after-another): 1) convection from fluid #1 to the tube wall, 2) conduction through the tube wall, and 3) convection from the tube wall to fluid #2. Let's apply this idea to a heat exchanger with two flowing fluids and a __rectangular__ solid wall between them (the left side of Figure 10.15). An example of this geometry is that of heat transfer across a single glass window pane on a winter day. Heat transfers from the air on the inside (warm side) of the window to the cold window pane. This would occur by conduction if the air were perfectly still, but the air is usually flowing because of such influences as the blower in normal heating systems and the normal movement of people inside the room. But motion also occurs because as the air near the cold window is cooled, it becomes more dense and, therefore, moves downward to displace the less dense warm air. The heat transferred to the window pane conducts through the glass to the other side, where it transfers to the cold outside air (again, mostly by convection).

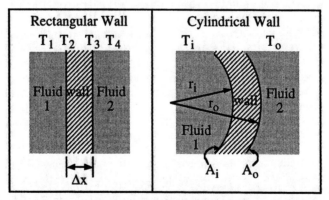

Figure 10.15 Heat transfer between two flowing fluids
separated by a rectangular or cylindrical wall

This system of a series of heat-transfer steps is analogous to multiple mass-transfer steps in series as discussed in Chapter 8. Similar to the mass-transfer case, we recognize that, at steady state, the rates of heat transfer through the steps must equal each other, so

$$\text{Transfer rate} = \dot{Q} = h_1 A(T_1 - T_2) = kA\frac{T_2 - T_3}{\Delta x} = h_2 A(T_3 - T_4) \qquad (10.25)$$

which can be solved to produce the heat-transfer equivalent of Equation 8.5, namely

$$\dot{Q} = \frac{T_1 - T_4}{\frac{1}{h_1 A} + \frac{\Delta x}{kA} + \frac{1}{h_2 A}} = \frac{\text{overall driving force}}{\sum \text{resistances}} \qquad (10.26)$$

One of the principles associated with Figure 10.15 and Equation 10.26 is the concept of a *limiting heat-transfer resistance*. For example, if the convection through Fluid 2 is very slow (h_2 is small and the resistance $1/h_2 A$ is large), all of the transfer will be limited by that slow rate. You can estimate the relative resistances of the three steps by simply comparing the values of the three resistances $1/h_1 A$, $\Delta x/kA$, and $1/h_2 A$. The relative values of these resistances can be compared to determine which, if any, is the limiting resistance.

Similarly, for a <u>cylindrical</u> wall (right side of Figure 10.15), Equation 10.26 becomes

$$\dot{Q} = \frac{T_i - T_o}{\frac{1}{h_i A_i} + \frac{\ln(r_o/r_i)}{2\pi k L} + \frac{1}{h_o A_o}} \qquad (10.27)$$

where L is the tube length, and A_i and A_o are the surface areas inside and outside the tubes, respectively. For Equations 10.26 and 10.27, you should be able to identify the resistances associated with convection through each fluid and for conduction through the wall.

For the sake of economy, we obviously want to maximize the heat transfer accomplished in a heat exchanger. From Equations 10.26 and 10.27, we can see that the transfer is increased as we decrease the resistances, particularly the limiting resistance as discussed above. Unlike the case of mass transfer, the conduction resistance (analogous to diffusion in mass transfer) is usually not the largest resistance. For example, for shell-and-tube devices, this is because the tubes are made from highly-conductive metal (with large k values) and have thin walls (the value of $\ln(r_o/r_i)$ is small). The resistances for the convective steps are also reduced by designing the exchangers so that there is significant turbulence in the flowing fluids (which increases the values of h_i and h_o). However, heat exchangers are prone to develop layers of residue ("*fouling*") on the tube walls with time, which provides a layer of low-conducting material (low k) and thereby increases the resistance to conduction through the tube walls. Therefore, maximizing heat-exchanger performance requires periodic cleaning of the exchangers, which is an expensive process involving the removal of the exchanger from service (often by shutting down the process) and the dissembling of the exchanger.

Equation 10.27 provides important insights into the contribution of heat-transfer area to the rate of heat transfer, but the estimation of the total heat-transfer area requirement is more complex. For example Equation 10.27 represents the conditions at a particular cross section of the exchanger, but the temperatures T_i and T_o change along the length of the device. Furthermore, the prediction of the values of h_i and h_o is complex. As a short-hand method of describing heat-exchanger performance, we use the *overall heat-transfer coefficient*, U_o, and an average temperature driving force, as presented in Equation 10.28.

$$\dot{Q}_{duty} = U_o A \, \Delta T_{avg} \qquad (10.28)$$

Typical units of U_o are $W/m^2 K$ and $Btu/hr \, ft^2 \, {}^\circ F$, where the temperature unit in the denominator represents a temperature difference and must be treated in the same way that the temperature difference is treated in the units for thermal conductivity and heat-transfer coefficient (see Figure 10.4). Table 10.5 provides some approximate values of U_o for some representative situations.

When detailed final designs are made, more accurate values of U_o are determined using methods that you will learn about in later classes.

Table 10.5 Approximate Values of U_o (from Reference 1)

Hot Stream: Cold Stream	U_o, Btu/hr ft^2 °F
Saturated vapor: Boiling liquid	250
Saturated vapor: Flowing liquid	150
Saturated vapor: Vapor	20
Liquid: Liquid	50
Liquid: Gas OR Gas: Liquid	20
Gas: Gas	10
Vapor (Partial condenser): Liquid	30

The term ΔT_{avg} in Equation 10.28 represents the temperature difference between the hot and cold streams averaged in some appropriate way over the entire heat exchanger. Various kinds of average temperature differences are used depending on the geometry of the exchanger. For single-pass exchangers, the appropriate form of ΔT_{avg} is the log-mean temperature difference, $\Delta T_{log\ mean}$ (often abbreviated $LMTD$), defined as

$$Log\ Mean\ Temperature\ Difference\ =\ \Delta T_{log\ mean}\ =\ \frac{\Delta T_1 - \Delta T_2}{\ln \frac{\Delta T_1}{\Delta T_2}} \qquad (10.29)$$

The subscripts in Equation 10.29 represent the two ends of the exchanger, so ΔT_1 and ΔT_2 are the differences between the hot and cold streams on the two ends, respectively, as defined in Figure 10.11. Thus, $\Delta T_{log\ mean}$ is simply a type of average driving force. Note also that ΔT_1 and ΔT_2 should have the same sign, either positive or negative. This is true because the temperature of the hot stream must be greater than that of the cold stream at both ends of the heat exchanger (as well as anywhere in the middle). Heat exchangers are typically designed with a minimum ΔT_1 or ΔT_2 of at least 5 °C.

For shell-and-tube exchangers, the inside area (A_i) of the tubes is smaller than the outside area (A_o), as shown in Figure 10.15. However, for this introductory discussion, we recognize that the design calculation represented by Equation 10.28 is approximate, and the small differences between A_i and A_o will be neglected. For the determination of heat-exchange area, assuming that $\Delta T_{avg} = \Delta T_{log\ mean}$, Equation 10.28 is rearranged to the form

$$A = \frac{\dot{Q}_{duty}}{U_o \Delta T_{log\ mean}} \qquad (10.30)$$

The use of this equation is illustrated in the following example:

Example 10.6

How much area is required for the countercurrent heat exchanger in Example 10.5?

Solution: Again, the diagram with the given information is drawn, where we have also labeled the two ends of the exchanger.

Saturated steam,   2   1   Saturated water,
280°F, $\dot{m}_{steam}$ 280°F, $\dot{m}_{steam}$

Oil, 110°F, 960 lb_m/min Oil, 35°F, 960 lb_m/min

From Equation 10.29,

$$\Delta T_1 = 280\text{-}35°F = 245°F, \quad \Delta T_2 = 280\text{-}110°F = 170°F$$

$$\Delta T_{log\ mean} = \frac{245 - 170}{\ln 245/170} = 205°F$$

In this case, the hot stream is a saturated vapor (steam), and the cold stream is a flowing liquid (oil). Thus, from Table 10.5, U_o is approximately 150 $Btu/hr\ ft^2 °F$, so Equation 10.30 becomes

$$A = \frac{53,280\ Btu/min}{\left(150\dfrac{Btu}{hr\ ft^2°F}\right)(205°F)}\left(\frac{60\ min}{hr}\right) = 104\ ft^2$$

The following table summarizes the procedures described in this section for sizing a heat exchanger.

Table 10.6 Basic Procedures for Preliminary Sizing of a Heat Exchanger

1. Use Equation 10.24 to determine the heat duty, $\dot{Q}_{duty}$, of the heat exchanger. To do this, use the "target stream" for which the inlet and outlet temperatures and flow rate are known.

2. Use $\dot{Q}_{duty}$ and Equation 10.24 to determine the unknown flow rate or temperature for the other side of the heat exchanger (additional information is needed if more than one of these is unknown).

3. Unless otherwise specified, assume a single-pass countercurrent heat exchanger and calculate $\Delta T_{log\ mean}$ as defined in Equation 10.29.

4. Determine the appropriate value for U_o from Table 10.5 and convert to the appropriate units.

5. Use Equation 10.30 to calculate the required heat-exchange area.

The Acid-Neutralization Problem

In the previous section, we learned that we need to remove heat in our acid-neutralization process. In this section, we learned a little about heat exchangers that can be used to accomplish such a task and how to apply energy balances and heat-transfer coefficients to the design of such devices. The logical conclusion is that we should add a heat exchanger to our process to cool the neutralized acid. Therefore, we need to apply the techniques discussed above to design a heat exchanger for accomplishing that task.

For the problem we are trying to solve, the required cooling is not intense, and the nearby lake has an abundance of cooling water. Therefore, the logical solution is to use a simple heat exchanger to cool the final product. <u>What flow rate of cooling water are we likely to need?</u> To answer this question, we need to determine the starting temperature of the lake water ($T_{cold,in}$) and the final temperature we would like that water to have ($T_{cold,out}$) and how much heat must be removed from the hot waste stream. From that information, Equation 10.24 can be used to determine the needed flow rate for the cooling water ($\dot{V}_{cold}$). For example, suppose that the lake water never exceeds 15°C (we want to design for the warmest lake water to provide a safe estimate) and that we want to prevent the cooling water from being heated to within 5 °C of the state-mandated maximum ($T_{cold,out} = 22°C$). Thus, summarizing our known temperatures and flow rates (see Figure 10.16):

Hot stream temperature in:	49.9°C
Hot stream temperature out:	27°C
Cold stream temperature in:	15°C
Cold stream temperature out:	22°C
Hot stream flow rate (approximate):	18,100 L/hr

Both the hot and cold streams can be considered to be water and to have the same values of density and heat capacity. At these temperatures,

Mean heat capacity:	4.17 kJ/kg °C
Density:	992 kg/m³

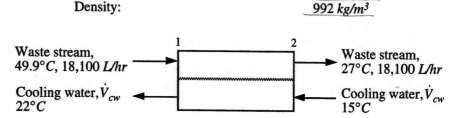

Figure 10.16 Temperatures and flow rates for the acid-neutralization problem

From Equation 10.24a and expressing the mass flow rate as the product of the volumetric flow rate and density,

$$\dot{Q}_{duty} = \left[\dot{V}\rho\overline{C}_p(T_{in} - T_{out})\right]_{hot}$$

$$= \left(18100\frac{L}{hr}\right)\left(992\frac{kg}{m^3}\right)\left(4.17\frac{kJ}{kg°C}\right)(49.9-27°C)\left(\frac{1\,m^3}{1000\,L}\right)\left(\frac{1000\,J}{kJ}\right)\left(\frac{1\,hr}{3600\,s}\right)\left(\frac{1\,W}{1\,J/s}\right)$$

$$= 477,000\ W$$

From Equation 10.24b $\qquad\qquad \left[\dot{V}\rho\overline{C}_p(T_{out} - T_{in})\right]_{cold} = \dot{Q}_{duty}$

or

$$\dot{V}_{cold} = \frac{\dot{Q}_{duty}}{\left[\rho\overline{C}_p(T_{out} - T_{in})\right]_{cold}}$$

$$= \frac{477,000\ W}{\left(992\frac{kg}{m^3}\right)\left(4.17\frac{kJ}{kg°C}\right)(22-15°C)}\left(\frac{1\,J/s}{W}\right)\left(\frac{1\,kJ}{1000\,J}\right)\left(\frac{3600\,s}{hr}\right)\left(\frac{1000\,L}{m^3}\right) = 59,200\frac{L}{hr}$$

Notice that this flow rate for the cooling water is only an average value, since it doesn't account for the fluctuations in acid flow rate from the manufacturing process or the corresponding fluctuations in the sodium hydroxide flow rate to maintain neutrality. However, the estimate is useful in helping us determine the approximate pump capacity and other flow parameters associated with the system which will be needed to provide the cooling water to the heat exchanger.

We now have the tools to calculate how much heat-transfer area such an exchanger must have. From Table 10.5 for two liquid streams, $U_o = 50$ *Btu/hr ft² °F*. Converting that value to metric units

$$U_o = 50 \frac{Btu}{hr\, ft^2\, {}^\circ F} \left(\frac{1055\, W}{Btu/s} \right) \left(\frac{1\, hr}{3600\, s} \right) \left(\frac{10.764\, ft^2}{m^2} \right) \left(\frac{1.8{}^\circ F}{{}^\circ C} \right) = 284 \frac{W}{m^2\, {}^\circ C}$$

Notice the method of converting the temperature difference from °F to °C. Also, for the countercurrent heat exchanger illustrated in Figure 10.16,

$$\Delta T_1 = 49.9 - 22 = 27.9{}^\circ C$$

$$\Delta T_2 = 27 - 15 = 12{}^\circ C$$

$$\Delta T_{log\ mean} = \frac{\Delta T_1 - \Delta T_2}{\ln \dfrac{\Delta T_1}{\Delta T_2}} = \frac{27.9 - 12}{\ln \dfrac{27.9}{12}} = 18.8{}^\circ C$$

So, from Equation 10.30,

$$A = \frac{\dot{Q}_{duty}}{U \Delta T_{log\ mean}} = \frac{477,000\, W}{\left(284 \dfrac{W}{m^2\, {}^\circ C} \right)(18.8{}^\circ C)} = 89.3\, m^2$$

We should add that heat exchanger to the process flow diagram, and the stream table should also include information about temperatures, as illustrated in Figure 10.17.

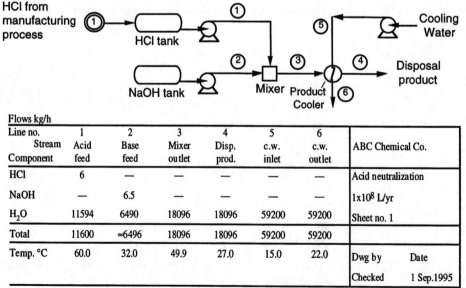

Flows kg/h

Line no.	1	2	3	4	5	6	ABC Chemical Co.
Stream	Acid feed	Base feed	Mixer outlet	Disp. prod.	c.w. inlet	c.w. outlet	
Component							
HCl	6	—	—	—	—	—	Acid neutralization
NaOH	—	6.5	—	—	—	—	1x10⁸ L/yr
H₂O	11594	6490	18096	18096	59200	59200	Sheet no. 1
Total	11600	≈6496	18096	18096	59200	59200	
Temp. °C	60.0	32.0	49.9	27.0	15.0	22.0	Dwg by Date
							Checked 1 Sep.1995

Figure 10.17. Process flow diagram for the acid neutralization process with heat exchanger

References

1. Incropera, F.P., and Dewitt, D.P., *Fundamentals of Heat and Mass Transfer*, 3rd ed., NY: John Wiley & Sons, 1990.

2. Douglas, J.M., *Conceptual Design of Chemical Processes*, NY: McGraw-Hill Book Co., 1988.

READING QUESTIONS:

1. A freshman chemical engineering student is walking across campus on a cold winter day. For each of the following scenarios, which forms of heat transfer (conduction, convection, and/or radiation) are operative and which form do you think is dominant? In each case, explain your answer.

    a. The student feels colder when a sudden gust of wind arises.

    b. After sitting down on a concrete bench, the student feels even colder, especially in that part of the anatomy in contact with the bench.

    c. Seeing some trash being burned in an incinerator, the student stands next to the fire and feels warmer.

2. To what extent does radiation heat transfer require the presence of a medium in order for the transfer to occur? How does this differ from conduction or convection heat transfer?

3. Name at least one example of conduction, convection, and radiation from everyday life (in addition to the examples mentioned in Reading Question 1).

4. Compare Equation 10.5 with Equation 8.1. How are they similar? How are they different?

5. Compare Equation 10.6 with Equation 8.4. How are they similar? How are they different?

6. What happens when heat is added to a saturated liquid? Does the addition of heat cause the temperature to change? Why or why not?

7. As he watches you boil potatoes on the stove, your neighbor suggests that if you turn up the flame on the stove, the potatoes will cook faster because the water will be "hotter." Is your neighbor correct? Why or why not?

8. Name 3 examples of heat exchangers which are part of your everyday life.

9. In a heat exchanger with one fluid carried inside a tube and another fluid outside the tube, what heat-transfer mechanisms operate in the following phases of transfer:

    a. transfer from the inside fluid to the tube wall

    b. transfer through the tube wall

    c. transfer from the tube wall to the outside fluid

HOMEWORK PROBLEMS:

1. Verify that Equation 10.23 follows from Equations 10.21 and 10.22 by performing the derivation using the assumptions summarized in the paragraph before Equation 10.22.

2. In the calculation to determine the outlet temperature of the acid-neutralization process (illustrated at the end of Section 10.2):

   a. rework the problem to find the outlet temperature assuming no heat is transferred to/from the surroundings ($\dot{Q}=0$, adiabatic) and using the following values:

   > inlet temperature of the HCl is $55\,°C$
   >
   > inlet temperature of the NaOH is $25\,°C$

   b. using the values in part a, determine how much heat would need to be removed (in units of *Watts*) in order to achieve the desired outlet temperature of $27\,°C$? Hint: remember that we are assuming that all streams have the same heat capacity and density as water (the heat capacity of water is $1.0\ cal/g\,°C$).

3. A stream of bread dough is fed steadily through an oven in a large bread-baking operation. The dough enters the oven at a rate of $110\ ft^3/hr$ and at a temperature of $70\,°F$. It exits the oven at $400\,°F$. Other properties of the dough include:

   > density $= 58.7\ lb_m/ft^3$
   >
   > heat capacity $= 0.9\ Btu/lb_m\,°F$

   At what rate is heat being added to the dough by the heating coils of the oven (in units of *Watts*)?

4. A liquid solution is passed steadily through an infrared radiative oven to sterilize it. The system is horizontal (no elevation change), and the velocities are small. Furthermore, the density of the liquid can be considered to be constant, and evaporation of the solution is negligible. The following are given:

Volumetric flow rate	$\dot{V}$	(same for inlet and outlet streams)
Density	ρ	(same for inlet and outlet streams)
Average Heat Capacity	$\overline{C}_p$	(same for inlet and outlet streams)
Reference Temperature	T_{ref}	(same for inlet and outlet streams)
Inlet Temperature	T_{in}	
Outlet Temperature	T_{out}	

   a. Derive an expression for the difference between the outlet temperature and the inlet temperature of the liquid.

   b. Does your solution make intuitive sense? In other words, does the radiant heat provided by the oven, the flow rate, and the heat capacity affect the temperature rise as you would expect?

5. Some of the most common sources of heat in a chemical plant are hot water and steam. The following questions explore the value of each of these as potential heat sources:

   a. Suppose that a hot water stream at 250°C is available as a heating source at a flow rate of 6.7 kg/s. How much heat can be extracted from this stream if it is cooled to 150°C? The heat capacity of water at these temperature is approximately 4.5 kJ/kg K.

   b. Suppose that saturated steam at 200°C is available as a heating source at a flow rate of 6.7 kg/s. How much heat can be extracted by condensing the steam to liquid water at the same temperature? (Note: $\Delta \hat{H}_{vap,200°C}$=1939 kJ/kg)

   c. Based on your answers from parts a and b, which of the above (hot water or steam) would you recommend as a source of heat? Why?

6. A liquid stream contains 27 mole% species A dissolved in a solvent. In a steady-state process, the stream enters a reactor where $2/3$ of the species A is converted to species B. The heat of reaction at the reactor temperature equals -918 kJ/kgmol of A reacted. How much heat (per kgmol of entering stream) must be transferred from the reactor to maintain a steady temperature in the reactor? Again, kinetic and potential energy effects can be neglected. Also, while there is some shaft work from the stirrer in the reactor, such work is usually negligible and can be considered so in this case. Hint: You will need to choose a basis.

7. A hot fluid flows inside a cylindrical pipe while a cold fluid flows along the outside of the pipe. Given the following data, what is the largest resistance to heat transfer?

           Hot fluid:
               Heat-transfer coefficient: 0.015 $W/cm^2°C$
               Density: 0.85 g/ml

           Cold fluid:
               Heat-transfer coefficient: 0.032 $W/cm^2°C$
               Density: 1.0 g/ml

           Pipe:
               Inside radius: 0.318 cm
               Outside radius: 0.476 cm
               Length: 50 cm
               Thermal conductivity: 0.53 W/cm°C

8. You have been assigned to design a heat exchanger to cool a gaseous stream (containing hydrogen and nitrogen) in an ammonia plant. The stream enters at 431 K, has a heat capacity of 3.45 J/g°C, and needs to be cooled to 402 K. The flow rate of the stream is 20 kg/s. Cooling water is available at 85 °F and has a specified maximum temperature of 120°F. The overall heat-transfer coefficient is approximately 570 W/m^2K. How much heat-exchange area is necessary to cool the process (gaseous) stream with the cooling water?

9. A heat exchanger is needed to heat a process stream from 300°F to 425°F. The heat will be supplied by steam on the shell side of the heat exchanger. Saturated steam is available at 600 psig (approx. 490°F, and $\Delta\hat{H}_{vap,600\,psig}=1699\ kJ/kg$). The heat capacity of the process stream is 3.0 kJ/kg°C. The process stream flows at a rate of 10 kg/s. The overall heat-transfer coefficient is approximately 150 Btu/hr ft² °F.

   a. What flow rate of steam (mass/time) is needed for this heat exchanger?

   b. The boiling point of water with which we are most familiar is 100°C. How is it possible to have saturated steam at 490°F?

   c. Estimate the required area of the heat exchanger in ft².

10. The last section of this chapter describes the sizing of a heat exchanger to cool the neutralized acid stream from 49.9°C to 27°C using cooling water available at 15°C. That calculation assumed an outlet temperature of 22°C for the cooling water. In contrast, this problem assumes that the outlet temperature of the cooling water can be as high as 27°C.

   a. Set up a spreadsheet to calculate the *volumetric flow rate of cooling water* needed and the *area of the heat exchanger required* to cool the outlet stream from the acid neutralization process. Perform the calculations for a range of cooling water outlet temperatures from 16 to 27°C in one-degree increments, assuming a countercurrent heat exchanger.

   b. Graph the results from part a. (cooling water flow and heat exchanger area as a function of cooling water outlet temperature) and provide a *brief qualitative explanation* of the observed behavior.

   c. Repeat the calculations and graphs from parts a. and b. for a co-current heat exchanger (caution: for this case, can the cooling water outlet temperature for this heat exchanger actually equal 27°C?).

   d. Based on the above calculations, which type of heat exchanger (countercurrent or co-current) would you recommend? Why?

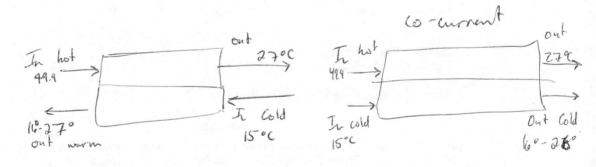

FACTORS:

Properties
Production
Price

CHAPTER 11

MATERIALS
(FROM WHAT SHALL WE BUILD THE EQUIPMENT?)

Up to this point in the book, we have been talking about the design of equipment to hold, convey, mix, react, and cool our hydrochloric acid stream, but we have not addressed the issue of which materials we might use to construct the equipment. The study of materials is very important, because modern processes frequently subject equipment to extremely harsh conditions for which common materials are not suitable. Significant effort is being invested in the development of new materials with greater strength and chemical resistance. This chapter will review the basic categories of materials and their general characteristics.

Metals and Corrosion

In the construction of chemical-processing equipment, metals have been the most commonly-used materials because of their great mechanical strength and because of humankind's long experience with metal refinement. The general features which define metals include their high ductility (ability to be shaped and formed), their high strength, and their density. Copper, aluminum, and iron are among the predominant metals used in equipment of all kinds. However, the metal which stands out because of its overwhelming popularity throughout modern history is an iron-based alloy called *steel.*

Steel is the name given to iron which contains a small amount of carbon. The relatively-weak iron is greatly strengthened because the carbon is present as single atoms or as tiny clusters of iron carbide within the regions of iron, and those atoms or clusters provide disruptions in the iron crystals which prevent the iron atoms from slipping past each other. The mechanical properties of the steel ultimately depend on 1) the amount of carbon in the steel (the higher the carbon content, the stronger and more brittle the steel), and 2) the heat treatment process to which the steel has been subjected (which affects the size of the carbide clusters, as well as the crystal structure and size of the iron crystals). Minor amounts of other metals are sometimes added to steel to give it special properties; the most common example is stainless steel, which is steel to which a small amount of chromium (and sometimes other metals) has been added.

One disadvantage of metals is that they can undergo *electrochemical corrosion*, in which the metal atoms are converted to metal ions, i.e. the metal is essentially dissolved. The general reaction is

$$M^o \rightarrow M^+ + e^-$$ (11.1)

The location where this occurs is called the *anode*, and the metal is said to be *oxidized*. Meanwhile, at another location, called the *cathode*, the electron which was released by the reaction in Equation 11.1 is consumed by some agent (i.e. that agent is *reduced*). Overall, the process can be represented as shown in Figure 11.1.

Properties: 1.) Tensile strength   4.) Ductility & elasticity
2.) Compressive strength   5.) Brittleness
3.) Shear strength   6) Stiffness

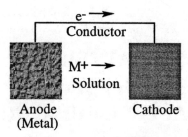

Figure 11.1 Elements of electrochemical corrosion

The four essential components of corrosion are represented in Figure 11.1, namely, an anode, a cathode, an ion conductor (some medium, usually an aqueous solution, in which the metal ions move away from the metal), and an electron conductor (some medium, usually a metal connection, through which the electrons travel from the anode to the cathode). For example, when a car is exposed to water (ion conductor) over a period of time, some of the metal (anode) undergoes corrosion, especially when that metal is in contact (electron conductor) with a less-reactive metal (cathode). Note that salt in the water increases corrosion by increasing the ability of the solution to conduct ions and by attacking protective oxide layers on metal surfaces.

The tendencies of various metals (and other chemical species) to participate in electrochemical corrosion can be represented in a chart such as in Figure 11.2. Such a chart is called an oxidation-reduction series, because it illustrates the relative tendencies for species to be oxidized (at the anode) or reduced (at the cathode). Thus, if a more noble metal is placed in electrical contact with a more reactive metal in a wet environment, the more reactive metal will

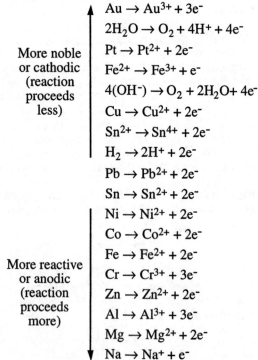

Figure 11.2 Abbreviated oxidation-reduction series

become the anode (will corrode). Additionally, the cathodic reaction (e.g. reduction of oxygen, reduction of hydrogen ions, plating of metal ions, etc) takes place on the surface of the more noble metal.

Corrosion can occur uniformly in a large metal surface, called *uniform corrosion*, or it can be very uneven. An example of uneven corrosion would be the *localized corrosion* in a small area causing a hole (or many tiny holes, *pitting*) to penetrate a pipe wall or tank wall.

Example 11.1

A piece of copper trim on the body of a car is in physical contact (metal-to-metal) with the steel panel of the car body. Which will corrode, the steel panel or the copper trim?

<u>Solution</u>: From the oxidation-reduction series in Figure 11.2, we see that copper is more noble or cathodic than iron (the principal component of steel). Thus, the iron will be more likely to undergo oxidation (corrosion).

In a chemical process, corrosion can be very hazardous because it may cause equipment to leak, releasing toxic chemicals into the environment and/or into the vicinity of human beings. In addition, it may cause equipment failure or lead to contamination of process streams. Therefore, equipment must be designed to prevent or compensate for the corrosion. Several strategies are employed to reduce corrosion, including the following:

•Similar metals: Eliminate the contact of dissimilar metals so that an anode and cathode are not already generated by the different potentials of two very different metals. If different metals must be present, prevent them from being in electrical contact.

•Thicker metal: When the corrosion is uniform, a common tactic is to simply use thicker metal in the construction to compensate for the anticipated corrosion. The thickness is selected so that the material will continue to perform its function for at least as long as the desired operational life of the equipment.

•Sacrificial anode: To prevent corrosion, a common tactic is to introduce a piece of another metal, a *sacrificial anode*, in contact with the metal to be protected. The sacrificial anode is composed of a metal which has a greater tendency to corrode than does the metal to be protected. As a result, the sacrificial anode corrodes instead of the protected metal. The sacrificial anode can then be replaced periodically to maintain the protection.

•Noble metal: Another protective measure is to choose a metal or metal alloy which has a lower tendency to undergo the corrosion reaction. This is the motivation behind the use of more "noble" metals, such as nickel or cobalt, in corrosive environments.

•Protective oxide: In some cases, it is possible to take advantage of the fact that certain reactive metals readily form a more stable and non-reactive "oxide" surface which reduces further corrosion. Chromium, the main additive in stainless steel, forms chromium oxide on the surface which resists the anode reaction of the iron. Similarly, aluminum, considered to be highly resistant to corrosion, is actually very reactive but quickly forms an oxide surface which is very protective against further reaction. Even iron can be induced to form a passive film in certain very oxidizing environments, and such a formation becomes part of the corrosion-inhibition strategy.

•Corrosion inhibitors: Some chemical agents, *corrosion inhibitors*, can be added to fluid systems to reduce corrosion. Because such agents are expensive, they are usually used in closed, circulating systems, where the need to replenish the agents is lower. Corrosion inhibitors act in a variety of ways. Some form a coating on the metal surfaces to reduce the contact between the metal and the corrosive environment. Others react with the small amounts of dissolved oxygen in the liquids, which is effective because the more common corrosion reactions require oxygen.

- •Paint: One of the most effective methods to prevent corrosion is to simply paint or coat the metal surfaces. This is practical as long as there is no danger of reaction between the paint and the process fluids or of other contamination of process fluids by paint residue.
- •Non-metals: Where feasible, replace metal with non-metals, such as ceramics or plastics. Examples of this practice are common, particularly the use of plastic pipe, tanks, etc.

Example 11.2

Nails used for construction must be strong and, therefore, are made mostly from steel. However, steel (iron) nails can corrode, thereby causing the building or other structure to weaken. To reduce the probability of this problem occurring, construction nails are *galvanized* (coated with zinc). Why would such a coating reduce the corrosion of the iron?

Solution: From the oxidation-reduction series in Figure 11.2, we see that zinc is more reactive or anodic than iron. Thus, the zinc will be more likely to undergo corrosion, thereby causing the iron to act as the cathode (i.e. the zinc will act as a *sacrificial anode*). As long as some of the zinc is still in contact with the iron, the iron will not corrode.

Corrosion is such an important and expensive consideration in many engineering applications that it is not uncommon to find engineers whose main expertise is the reduction of corrosion.

Ceramics

Ceramics are an important class of materials because of their high resistance to reaction, even under very harsh conditions. Ceramics are chemical compounds which involve the ionic and covalent bonding of atoms in crystalline structures. Examples include alumina (Al_2O_3), silica (SiO_2), and diamond (a lattice of carbon). Because of the ionic and covalent bonding in ceramics, the electrons are locked into the bonds and don't exchange readily. This high degree of electron "locking" results in very low electrical conductivity and also contributes to the very low reactivity of ceramics, even at high temperatures. Ceramics are also relatively hard and brittle, because the atoms do not exchange bonds easily (which would be necessary for a material to deform without fracture).

An important material related to ceramics is glass. Glass is composed of silica chains in a non-crystalline structure, thereby disqualifying glass from being a ceramic. Glass is not only non-crystalline, but it is a supercooled liquid which flows, albeit very slowly (very old glass windows are measurably thicker at the bottom than at the top). Though glass is non-ceramic, the chemical properties of glass resemble those of ceramics to such a degree that glass is usually included in the discussion of ceramics as a "pseudo-ceramic." Some very expensive glass has been treated in a way to produce higher crystallinity, thus causing the material to be more like quartz, which is the purely-crystalline analog of glass and is a true ceramic.

In chemical engineering, ceramics are used for applications where high-temperature resistance and low reactivity are needed. Such applications include the linings of furnaces, support material for catalysts, packing for columns, linings of pipes and tanks carrying highly-reactive materials, and applications where electrochemical corrosion of metals would be a problem. Because of its transparency, quartz is used as a window for high-temperature reactors. Finally, brick, concrete and mortar are composed of ceramic compounds and provide structural materials for a wide variety of applications.

Polymers

In the past fifty years, polymers have been developed which provide a wide variety of properties. They are lightweight compared with metals or ceramics and are resistant to electrochemical corrosion. Furthermore, they can be mechanically flexible and tough. Because of the nearly-infinite number of possible polymer formulations, the properties and characteristics of polymers are likewise nearly infinite.

The word "polymer" means "many pieces," which reflects the fact that polymers are composed of many repetitions of a particular atomic sequence. Most polymers are based on a backbone chain of carbon atoms, to which various atoms are attached. The simplest polymer is polyethylene, which has the structure illustrated in Figure 11.3.

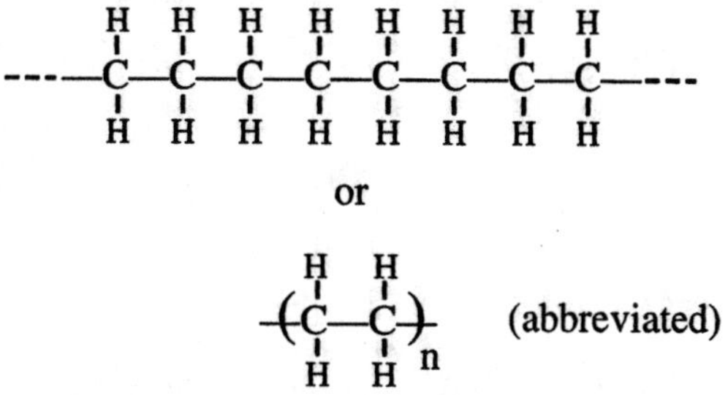

Figure 11.3 Structure of polyethylene

The drawing in the top part of Figure 11.3 represents a very long chain of carbon atoms with hydrogen atoms attached. The smaller drawing in the lower part of the figure is an abbreviation of the upper structure, where the atoms in parentheses represents the portion of the molecule which is repeated many times (the repeating unit). The "n" to the lower right of the parenthesis signifies that there are many of those repeating units. Polyethylene is a very common polymer, which is used to make items such as garbage bags and sandwich bags.

Different polymer properties are developed by substituting other atoms in place of the hydrogen atoms of polyethylene. For example, a common polymer is polyvinylchloride (PVC), which is the polyethylene molecule with one hydrogen atom out of four replaced by a chlorine atom (Figure 11.4). PVC is used for many applications, including such things as sprinkler pipe and surgical tubing. A polymer which is very resistant to chemical attack is polytetrafluoroethylene (trade name Teflon™), which is similar to polyethylene except that it has

polyvinylchloride   polytetrafluoroethylene

Figure 11.4 Two common polymers similar to polyethylene but with hydrogen substitution

fluorine atoms bound to the carbon in place of the hydrogen atoms (Figure 11.4). Because of its resistance to reaction, Teflon is used for such things as coatings on cooking pans and utensils. Much larger chemical groups can also be substituted for the hydrogen atoms, and the number of additional variations is nearly endless.

The properties of a polymer can also be varied by substituting other atoms or groups in place of the carbon atoms in the backbone of the chain. Single carbon-carbon bonds, such as those represented in Figures 11.3 and 11.4, rotate easily and result in relative flexibility in the polymer. However, backbone chains can also contain carbon-carbon double bonds, carbon-carbon triple bonds, rings, and other rigid structures which restrict chain rotation and make the polymer less flexible (see Figure 11.5). Represented in Figure 11.5 are polybutadiene (synthetic rubber) and polycarbonate (used in the windows of the space shuttles).

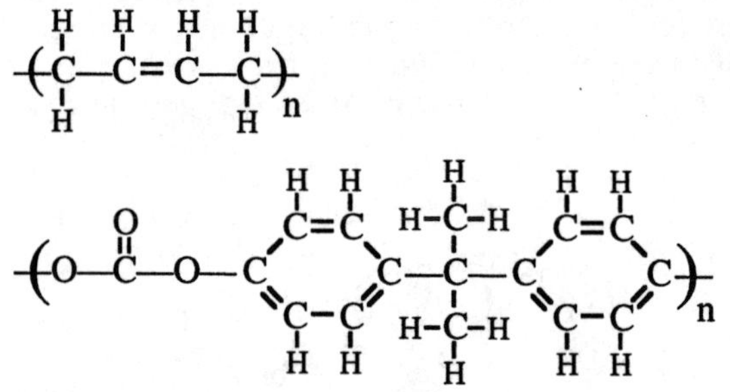

Figure 11.5 Two polymers with less-flexible structures in the polymer backbone

Polymer properties are also affected by the interactions between polymer chains. Those interactions can be affected by the following:

- •Crosslinking: Covalent bonding often occurs between chains to link the different polymer chains to each other. This crosslinking can vary in terms of the length and the frequency of the crosslinks.
- •Weaker bonding: Weak chemical interactions also occur between polymer chains, as various chemical groups on one chain attract groups on another chain. Van der Waals attractions and hydrogen bonding are two names given to these kinds of attractions.
- •Physical tangling: Chains or parts of chains can tangle with neighboring chains to restrict chain movement. This is influenced by the average chain length, since short chains can more easily move around each other to provide great fluidity or flexibility (e.g. flexible polyethylene garbage bags), while long polymer chains entangle with each other and cause a polymer to be very viscous or rigid. The tendency of polymer molecules to entangle is also affected by the frequency and length of side chains, chains which branch off of the main chain.
- •Crystallinity: All of the interactions mentioned above, along with the nature of the polymer chains themselves, affect the ability of the polymer chains to align themselves in rigid, dense, well-ordered crystalline structures. Many polymers have regions that are crystalline in this sense, and the physical properties of such polymers are influenced by the percentage of the polymer which is found in such a crystalline structure.

The field of polymer science is a dramatic example of using chemistry and chemical processing to make products which have a positive impact on our lives. Such products range from contact lenses to tennis shoes to space ships. Polymers are also used in chemical processing equipment, mostly because of their immunity to electrochemical corrosion and to the resistance of some polymers to attack by certain chemicals. Thus, tanks, pipes, and pumps made of polymers are used in corrosive environments. Where the strength of metals is needed, the polymers are sometimes used as a lining in metal pipes or tanks or as a coating around metal pump parts.

It is evident from the above discussion that a knowledge of carbon chemistry is critical to polymer processing and to the use of polymers as structural materials in chemical process equipment. *Your organic chemistry classes will help provide you with the background needed to work in this exciting subfield of chemical engineering.*

Composites

In recent years, the need for lighter, stronger, chemically-inert materials has led to the development of composites, materials comprised of a combination of polymers and either metals or ceramics. A common form of composites is a polymer in which are embedded many fine fibers of metal, graphite, or glass. The oldest example of this kind of composite is "fiberglass," which has tiny glass fibers embedded in a polymer. Sometimes the embedded glass or metal is woven into a laminate, and the composite contains layers of the laminate alternating with layers of the polymer. The impact of composites on our lives is larger than we normally realize, with such products as tennis rackets, airplane wings, and circuit boards comprising just a few of many examples.

Strength of Materials

For all of the materials discussed above, it is important to describe the strength of the material. Strength is commonly characterized as the force needed to stretch a standard sample of the material by using a large tensile testing machine (Figure 11.6). The force (measured by the tensile testing machine) is expressed as a *stress* (σ_x, defined as the force per original cross-sectional area) and the amount of stretch as a *strain* (ε_x, increase in length divided by the original length). As the force is plotted versus time, a stress-strain curve is obtained (Figure 11.7).

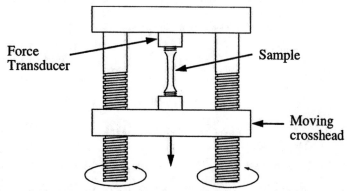

Figure 11.6 Tensile testing machine for characterizing the strength of a sample

For most materials, the stress-strain curve will exhibit the regions depicted in Figure 11.7, although the relative sizes of those regions varies significantly between types of materials. As the material first begins to stretch, it will usually undergo *elastic deformation*, that is deformation which is linearly proportional to the stress (force) applied and which will completely recover (return to its original dimensions) when the stress is removed. The slope of that linear portion of the stress-strain curve is called the *modulus of elasticity* (E_m), which has the same units as stress

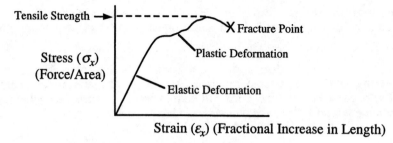

Figure 11.7 A typical stress-strain curve

(e.g. *psi*). In other words,

$$E_m = \frac{\sigma_x}{\varepsilon_x}$$ (11.2)

As the sample is stretched further, most materials will begin to undergo *plastic deformation*, which is when the material is permanently deformed and will not return to its original shape. Finally, most materials will reach a *fracture point* where the sample breaks. The highest stress sustained by the material is called the *tensile strength* and is the most common indicator of the material strength. It is important to note that the magnitude of the strain corresponding to these points in the stress-strain curve is relatively small (a few percent) for most materials, especially for rigid materials like metals or ceramics. In other words, rigid materials stretch very little, as you would expect.

Materials are commonly characterized by such parameters as the modulus of elasticity, the point at which plastic deformation occurs, and the tensile strength. Table 11.1 provides some comparative values for a few materials.

Table 11.1 Physical Properties of Some Materials

Material	Density (g/cm³)	Modulus of Elasticity (psi)	Tensile Strength (psi)
Iron	7.9	15×10^6	$42\text{-}73 \times 10^3$
Steel	7.9	30×10^6	$47\text{-}200 \times 10^3$
Copper	9.0	15×10^6	$25\text{-}50 \times 10^3$
Aluminum	2.7	10×10^6	$10\text{-}19 \times 10^3$
Magnesium	1.7	6.5×10^6	$23\text{-}50 \times 10^3$
Glass Fiber	2.5	10×10^6	250×10^3
Clay brick	2.5-5	15×10^6	$16\text{-}20 \times 10^3$
Polystyrene	1.1	4.5×10^5	7×10^3
Polyethylene	.95	$.2\text{-}1.2 \times 10^5$	$2\text{-}4 \times 10^3$

The use of Equation 11.2 is illustrated in Example 11.3.

Example 11.3

How much force would be required to elastically stretch a rod of aluminum (diameter = 0.44 in) from a length of 4.0 cm to 4.004 cm?

Solution: The cross-sectional area is $A = \frac{\pi}{4}d^2 = \frac{\pi}{4}(0.44\ in)^2 = 0.152\ in^2$, and the strain is

$$\varepsilon_x = \frac{\Delta l}{l_o} = \frac{0.004\ cm}{4.0\ cm} = 0.001$$

Also, from Table 11.1, $E_m = 10 \times 10^6\ psi$

So, Equation 11.2 tells us that

$$\sigma_x = \frac{F}{A} = E_m \varepsilon_x$$

or $F = AE_m\varepsilon_x = (0.152\ in^2)(10 \times 10^6\ psi\ )(0.001) = 1,520\ lb_f$

With this very brief introduction to materials, you should have a few specific skills and a broad perspective. In terms of specific skills, you should be able to predict which of a pair of metals in contact in a wet environment will preferentially corrode. You should also be able to calculate the stress or strain for elastic deformation of a material. More broadly, you should appreciate the importance and diversity of modern materials in our everyday lives and in industrial chemical processing.

READING QUESTIONS:

1. Which kind of steel would be the best candidate for a structural beam where a great deal of strength is needed: low-carbon steel or high-carbon steel? Why?

2. One common tactic to prevent the corrosion of steel pipelines is to attach blocks of zinc or magnesium to the pipeline. How do these blocks protect the pipeline?

3. From what you have read, why would brick be a popular outer cover for homes in geographical regions where the weather is harsh?

4. In a process to make a polymer, you can change the average chain length of the polymer by varying the reactor conditions in the process. After some trial runs, you find that the polymer initially produced does not have sufficient strength for the intended application. Which way should you change the chain length: shorter or longer? Why?

HOMEWORK PROBLEMS:

1. For each of the following pairs of metals in contact in an aqueous solution, which metal will corrode?
    a) iron and zinc
    b) iron and copper
    c) gold and copper
    d) iron and tin

2. The structural beams of an offshore drilling rig are made of steel and are partially submerged in sea water. The submerged portions are observed to corrode, and your engineering assignment is to decrease the corrosion rate of the steel by putting a second metal in contact with the beams. For each of the following arrangements, indicate whether it will decrease the corrosion of the steel beams and explain your answer

   a. Zinc placed above the water line (not submerged)

   b. Tin placed below the water line (submerged)

   c. Lead placed below the water line (submerged)

   d. Nickel placed above the water line (not submerged)

   e. Magnesium placed below the water line (submerged)

3. Steel has a modulus of elasticity of approximately 30×10^6 *psi*. A sample of steel originally 2.0 *inches* long and with a cross-sectional area of 0.196 *in²* is elastically stretched using 10,000 *lb_f*. How many *inches* beyond the original length will the sample stretch?

4. Various lightweight composite materials are being tested for possible use as structural columns in a space station. Small samples of each composite (cross sectional area = A_{sample} in units of *in²*) are stretched elastically in a large press until the sample length has increased from 2 *cm* to 2.03 *cm*, and the stretching force applied by the press (F_{press} in *lb_f*) is recorded at that point. The Space Agency wants to know how much force (F_{column} in *lb_f*) would be required to elastically stretch a structural column (cross sectional area = A_{column} in units of *in²*) by 1.2%.

   a. Derive a relationship for calculating F_{column} from F_{press}, A_{sample}, A_{column}, and the given magnitudes of elastic stretching.

   b. Help prepare the Excel™ spreadsheet section shown below to perform the calculation in part a. using data entered into the other cells as shown. Write the function exactly as it would be entered into cell B8 to provide the value of F_{column} for any values entered into cells B3, B4, and B6.

	A	B	C
1	Prediction of Force on Structural Column		
2			
3	Force on Sample to stretch 2->2.03 cm (lbf):	30,000	
4	Cross-sectional Area of Sample (in^2):	0.196	
5			
6	Cross-sectional area of Structural Column (in^2):	40	
7			
8	Force on Column for 1.2% Stretch	function you enter	
9			

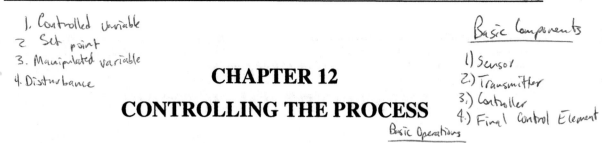

1. Controlled variable
2 Set point
3. Manipulated variable
4. Disturbance

Basic Components

1) Sensor
2.) Transmitter
3.) Controller
4.) Final Control Element

Basic Operations
1.) Measurement
2.) Decision
3.) Action

CHAPTER 12

CONTROLLING THE PROCESS

Section 12.1   Strategies of Process Control

Up to this point, the description of the acid-neutralization problem has used an average flow rate and concentration of HCl and, from these, the flow rate and concentration of NaOH needed to neutralize the HCl were determined. If the HCl flow rate and concentration and the NaOH concentration all remained perfectly constant, the NaOH flow rate could likewise be kept constant. However, process variables rarely remain constant on their own. Rather, they tend to vary with time as illustrated in Figure 12.1. In Figure 12.1a, the process variable of interest (e.g. a flow rate) changes significantly with time but continues to oscillate around the desired value. Substantial changes such as these make it difficult to obtain a product with the desired characteristics (e.g. concentration) at any given time. Even worse, if the product shown in Figure 12.1a is used as an input to a downstream process, the quality of that downstream process is severely compromised. Figure 12.1b illustrates the undesirable situation where the process steadily drifts away from the desired output. In contrast, Figure 12.1c illustrates the desired situation where any variations in the product characteristics are kept within a narrow band around the desired value. In order to achieve this, we need a method of controlling the process.

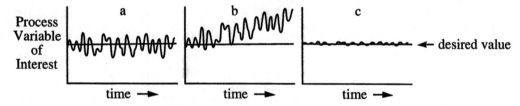

Figure 12.1   Examples of variation in the value of a process variable where that value
a) oscillates around the desired value, b) drifts away from the desired
value, and c) stays close to its desired value

The HCl entering the acid-neutralization process comes from the existing part of the plant and is received into a holding tank (Figure 12.2). It is then drawn from that tank via a pump, which delivers it to the acid-neutralization process. Furthermore, our experience shows that the production rate of that byproduct fluctuates and also drifts as depicted in Figure 12.1a and 12.1b. If the production rate exceeds the flow rate of the pump for a sufficiently-long time, the HCl will overfill the tank, and if the production rate falls below the flow rate of the pump, the tank will run dry. Thus, the flow rate of HCl leaving the holding tank would need to be adjusted

Figure 12.2 The holding tank for HCl at the beginning of the acid-neutralization process

frequently to avoid these problems, and the most common way of doing this is to place a valve downstream of the centrifugal pump (Figure 12.3). Further, with this variation in HCl flow to the acid-neutralization process, the NaOH flow rate would need to be constantly adjusted in order to prevent the pH of the stream discharged to the lake from falling outside the required limits. In addition, the amount of heat removed from the neutralized stream would need to be adjusted in order to obtain the desired outlet temperature. Clearly, several controls are needed to successfully run the process.

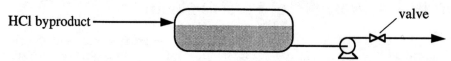

Figure 12.3 The use of a valve to vary the flow rate from the HCl holding tank

Many years ago, processes were controlled manually by people, called *operators*, who watched the dials and gauges which displayed what was going on in the process. From the readings on the dials and gauges and based on their experience and training, the operators turned knobs and valves to make corrections in the process. A primitive example of manual control applied to the liquid level in the HCl holding tank would be to have an operator frequently walk out to the tank and look at its liquid level and then manually open or close the valve on the outlet stream if the liquid level in the tank were too high or too low. This form of control could be improved by adding a measurement device or sensor to detect the liquid level and to transmit that level via an electronic signal to the control room so that it could be determined remotely without having to walk out and look into the tank. Also, a pneumatic valve (operated using high-pressure air) could be added to allow remote adjustment of the flow rate out of the tank. With these additions, the operator could determine the level in the tank and manually adjust the outlet flow to compensate for level changes without ever leaving the control room.

An even better alternative to the manual control discussed above is a fully-automated control system. In such a system, a controller (usually a computer) performs the role of the operator (Figure 12.4). Thus, the signal from the level sensor is sent to the controller instead of the operator. The controller then sends an output signal directly to the valve in order to make the proper adjustment in the flow rate. Such automatic control systems consisting of hardware and software operate continuously without direct human involvement. They also operate precisely and consistently. Because of this, they provide more precise control and improved safety by avoiding human error and variability.

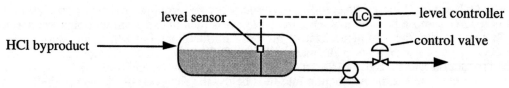

Figure 12.4 Automated control of the level in the HCl holding tank

Strategies have been pursued for many years to perfect these automated systems through the development of improved control algorithms (software) and better hardware. Process control is now an essential element of modern chemical processes and has resulted in enhanced product quality and increased safety. In fact, process control is considered as a standard subdiscipline of chemical engineering and represents an area of specialty for many chemical engineers. The purpose of this section is to introduce you to this aspect of chemical engineering by describing a couple of basic automatic control strategies which are widely used.

Let's begin our discussion with a definition:

Process control is the adjustment of process variables in response to changes in other parameters for the purpose of maintaining one or more variables within desired specifications.

To illustrate, we will consider holding a particular variable, called the *controlled variable*, at a desired value, called the *setpoint*. Variables which affect the value of the controlled variable are called *input variables* or *inputs*. Some of those inputs undergo uncontrolled changes, called *disturbances*, to cause a difference between the controlled variable and the setpoint (that difference is called an *error*). Typically, the controller is used to adjust one of the inputs, the *manipulated variable*, to reduce the error. In order to better understand these terms, let's see how they apply to the HCl holding tank in the acid-neutralization process. As mentioned previously, a control system is needed to keep this storage tank from overfilling or draining completely. Therefore, the controlled variable is the liquid level in the tank. The height at which we want the liquid to remain (e.g. 2 *ft*. from the top of the tank) is the setpoint. Variables which affect the liquid level, considered inputs from a control perspective, include the flow rate of HCl into the tank and the flow rate of HCl out of the tank. Changes in the flow rate of HCl into the tank represent disturbances, since that flow rate is uncontrolled and determined by some unknown conditions upstream. The manipulated variable is the flow rate of HCl out of the tank which is adjusted in order to keep the proper level of liquid in the tank. With this background, two common control strategies will now be defined and discussed.

Feedback Control

In Feedback Control, the <u>controlled variable is monitored</u> as a function of time, and the response of the control system is based on the measured values of the controlled variable. For example, measurements of the level (controlled variable) in the acid holding tank are fed back to a level controller (LC), which is a device (typically a computer) designed to manage the control operation and which uses that information to adjust the flow rate of the stream leaving the tank (manipulated variable) as shown in Figure 12.4. The structure consisting of the sensor, controller, and control valve, together with the communication lines which connect them (electronic or pneumatic), is commonly referred to as a *control loop*.

Feedback control can be applied to at least two other areas of the acid-neutralization process. For example, it can be used to control the pH of the neutralized acid stream, where the controlled variable, the pH of the liquid leaving the mixer, would be measured by a pH sensor and fed back to a concentration controller (CC) which adjusts the NaOH flow rate (Figure 12.5a). Feedback control could also be used to control the final outlet temperature of the neutralized acid by sensing that temperature and adjusting the cooling water flow rate (Figure 12.5b).

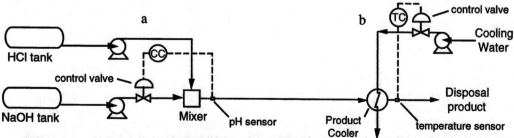

Figure 12.5 Feedback control of a) the pH and b) the temperature of the disposal product

To illustrate, let's examine the control loop for controlling the pH. Incidentally, pH is a difficult variable to control because it can change dramatically with small changes in acid or base concentration (as you may have learned from pH titrations in the chemistry laboratory). Suppose that the flow rate of HCl into the process changes (a disturbance), causing the pH coming from the mixer to vary from its setpoint. The controller subtracts the reading of the pH sensor from the specified setpoint to determine the error. The controller then uses this error to calculate an output signal which is sent to the control valve controlling the NaOH flow rate. The position of the valve is then adjusted according to this signal in order to change the NaOH flow rate and drive the pH toward its setpoint. Note that, in feedback control, the value of the controlled variable must be different from the setpoint in order for the controller to act. Furthermore, as the pH approaches the setpoint, the controller will continue to make adjustments in response to the updated value of the error. When the corrections are of the proper magnitude, the pH will approach its desired value as quickly as possible given the system dynamics (Figure 12.6a).

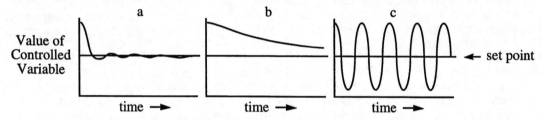

Figure 12.6   Feedback control response curves when the corrections are a) of the proper
             magnitude, b) too small, and c) too large

The engineer can change the response of a control loop by telling the controller how much of a change to make in the output signal for a given error. The engineer does this by changing the parameters in the controller. If the changes specified for a given error are too small, the response of the controller will be slow or sluggish as shown in Figure 12.6b. However, if the magnitude of the specified change is too large, the controller will overcorrect causing oscillating behavior (Figure 12.6c) or perhaps even an unstable response (where some process variables even reach a dangerous level). Later in your chemical engineering curriculum you will learn to describe the time-dependent behavior of chemical processes (your calculus becomes particularly important here).   You will also learn about the algorithms used by controllers to determine the appropriate output signal, and how to determine the parameters which are best for a given control situation.

Feedforward Control

Feedforward control operates by <u>measuring the disturbance(s)</u>, predicting the effect of the disturbances on the controlled variable, and making corrections to offset the predicted effect. In contrast to feedback control, the controlled variable is not measured in feedforward control. The error prediction is based solely on how the process is expected to behave (from a mathematical model), and the controller determines the needed correction from that prediction.

To illustrate feedforward control, let's apply it to the control of the pH of the mixer outlet to compensate for variations in the HCl flow rate entering the acid-neutralization process. Feedforward control relies on measuring the disturbance(s); therefore, we would need to add a flow sensor in the stream from the HCl holding tank. The measurements from this sensor would be sent to a feedforward controller which would predict the effect of changes in this measured variable on the final pH and would determine the output signal needed to adjust the NaOH flow rate in order to achieve the desired pH (Figure 12.7). For example, if the HCl flow rate from the holding tank increased, the feedforward controller would immediately increase the NaOH flow rate, without waiting for a change in the final pH to occur before responding.

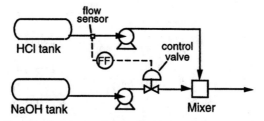

Figure 12.7 Feedforward control of the pH in response to variations in HCl flow

This illustration of feedforward control to manage the pH was based on only one disturbance, namely variations in HCl flow rate, but other disturbances might also be anticipated. For example, the inlet concentration of the HCl might vary. The concentration of the NaOH solution might also vary, even though it is purchased commercially (even small variations could be important because of the sensitivity of the final pH to very small changes). How would you construct feedforward control loops to compensate for these variations?

Comparison of Strategies

Each of the two control strategies, feedback and feedforward control, has advantages and disadvantages. Feedback control actually measures the controlled variable and is therefore capable of responding to all disturbances which affect this variable. It has the disadvantage, however, that the controlled variable must deviate from the desired value before corrective action is taken. Thus, it is possible that systems which respond slowly to corrections (i.e. systems which require a long time to come back to the set point after a deviation occurs) may deviate significantly from the desired setpoint when under feedback control. Feedforward control has the advantage that it can correct for measured disturbances before the controlled variable is affected. Therefore, it has the potential of preventing the controlled variable from deviating from the desired value and can be very advantageous in some systems. However, feedforward control does not directly measure the controlled variable, but relies on some kind of model to predict its value from the values and trends of the input variables. Any errors or imperfections in the model, such as inaccurate predictive equations or not accounting for all disturbances, will lead to deviations of the controlled variable from the desired value, a situation which cannot be corrected with feedforward control. Consequently, feedback control is the control strategy most commonly used. In situations where feedforward control is particularly advantageous, it is usually used in combination with feedback control to achieve the best of both strategies.

In keeping with the philosophy described above, Figure 12.8 illustrates the three feedback control loops we have mentioned for the acid-neutralization process (the level in the HCl holding tank, the pH of the neutralized acid, and the cooler outlet temperature). It also shows the one feedforward loop we have discussed for controlling the pH of the neutralized acid used in combination with the feedback loop for that same system.

Years ago, automatic control was accomplished exclusively by pneumatic devices (devices that operated using high-pressure air), which were used to convey the signals from the sensing equipment, to act as the process controllers, and to apply the corrections from the controllers to the process devices which affected the input variables. The more-recent development of digital

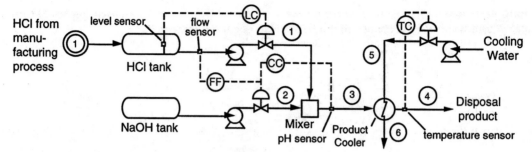

Figure 12.8 The acid-neutralization process with control loops, including a level
controller (LC), temperature controller (TC), concentration controller (CC),
and feedforward controller (FF)

computers as process controllers has provided much greater flexibility in control equations and
has added the ability to interface multiple control systems together. That development has made
it necessary for equipment to interface with computers, as discussed in the next section.

Section 12.2   How Do Computers Talk to Equipment?

As mentioned in the previous section, modern control of process equipment is accomplished
with the use of computers. In this role, computers must be able to "see" what is happening in a
process and then to "tell" the process how to change. This requires that the computer be able to
"talk to" the process equipment. Hence, it must be able to read a signal from the equipment in
such a way that the value of the signal (magnitude, frequency, etc.) becomes an *input* into a
computer program. The computer must also be able to generate a response which it can send out
to the equipment as an *output* from the computer program. These inputs and outputs are
electrical signals which interface with the electronic components that are part of the process
equipment.

There are two types of electrical signals which are typically used for communication between
electronic devices, a <u>current signal</u> and a <u>voltage signal</u>. *Current* refers to the number of
electrons per second which flow along an electrically-conductive material, such as a wire, and is
usually expressed in units of *amperes* or "*amps*." A high current corresponds to the flow of a
large number of electrons, whereas a small current corresponds to a relatively small number of
flowing electrons. The electric potential or *voltage* is a measure of the energy of the electrons
and represents the driving force for current flow. Current moves in a conductor from a location
where the voltage is high to a point where the voltage is lower. In addition, the material in which
the current is moving provides a *resistance* to the flow of electrons. Because of the high
mobility of electrons in metals, the electrical resistance of metals is much lower than that of most
other materials. Therefore, the conducting material of choice is usually a metal, although other
materials such as semiconductors or superconductors are used when dictated by the specific
application.

An analogy to fluid flow in a horizontal pipe (see Example 7.5) can be used to help us
understand these quantities. Measurement of the current is analogous to measuring the flow rate
of the fluid through the pipe. Further, the voltage provides the driving force for current flow in
much the same way that pressure provides the driving force for fluid flow. Therefore,
measurement of the voltage signal is analogous to measurement of the pressure drop in the pipe.

Finally, the electrical resistance in analogous to friction in the pipe. The relationship between current, voltage, and resistance is expressed by Ohm's Law, which is:

$$V = IR \qquad (12.1)$$

where V represents the voltage difference between two locations in a material, I is the current passing through that material between those locations, and R equals the resistance of the material between those locations. Thus, for systems where the resistance is fixed, the voltage and current will vary together. However, for some systems, the electronics are configured so that the voltage signal is the primary signal. In other systems, the variation in current carries the desired information.

To provide a standard communication system for computer-equipment communication, two systems of electrical signals have been developed. They are a current signal ranging from 4 to 20 *milliamperes* (1 *milliampere* = 0.001 *ampere*) and a voltage signal ranging either from -5 to +5 *volts* or from 0 to 10 *volts*. Hardware for interfacing between computers and industrial equipment accepts one or both of these types of signals, and equipment designed to be connected with computers is designed to generate one or both of these kinds of signals.

Electronic equipment can be used to <u>measure</u> physical quantities if the magnitude of the physical quantities can be converted to electrical voltage or current, i.e. to an electrical signal. The device which converts the physical measurement to an electrical signal is called a *transducer*, and transducers exist for many types of measurements, including pressure, temperature, pH, flow rate, liquid level, etc.

Equipment can also be <u>controlled</u> by a computer. Devices such as pumps, valves, heating devices, etc. can be designed so that a current signal (4-20 *mA*) or voltage signal (0-10 *V*) from a computer will affect how fast the pump operates, how far the valve closes, how much heat the heating device delivers, etc.

An important part of the communication between computers and equipment is the conversion between analog and digital information. The kinds of electrical signals we have been talking about, like most physical parameters, are usually *analog*, meaning that they vary by changing smoothly over a range containing an infinite number of values. By contrast, computers represent values as numbers from a finite set of discrete numbers. In other words, they see values as combinations of digits (one, two, three, etc.) and, therefore, deal with *digital* signals. Digital signals are represented by discrete numbers and undergo changes which are multiples of the minimum difference between integers (Figure 12.9).

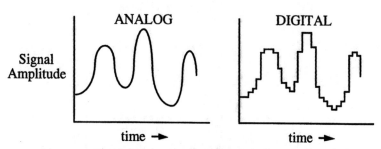

Figure 12.9 Comparison of analog and digital signals

If a signal is to be processed by a computer, it must be changed from a continuous stream of varying electrical current or voltage to a string of digital values that the computer understands.

To accomplish this task, hardware has been developed which converts analog signals to digital binary numbers. These *analog-to-digital converters* (A-to-D converters or A/D converters) are usually in the form of computer cards which plug into special slots inside the computer. The wires carrying the current or voltage from a transducer system are connected to the card, and the electronics on the card "sample" the signal (assess its magnitude), convert that magnitude to a digital value, and store that value in a particular "memory" location on the card. At a preset frequency, the card repeats the process and replaces the value in the memory with an updated value. The computer can be programmed to read the value in that memory of the card as often as desired and to use it as necessary.

Similarly, if a computer is to control or communicate with equipment, values from the computer (originally in digital form) may need to be changed from strings of digital values to a stream of electrical current or voltage. To accomplish this task, hardware has been developed which converts digital numbers to either electrical current or voltage. These *digital-to-analog converters* (D-to-A converters or D/A converters) are also in the form of computer cards which plug into special slots inside computers. The wires carrying the current or voltage to some electronic equipment are connected to the card, and the card repeatedly changes digital values stored in a memory on the card into current or voltage signals. Also, the computer is programmed to repeatedly send values to that memory location on the card.

To further understand the communication between computers and equipment, it is important to understand that the digital number system used by computers is a *binary* number system, i.e. it represents numbers using a combination of "ones" and "zeroes" in a 2-based number system. This is because computer operation and communication rely on the presence ("one") or absence ("zero") of some electrical/magnetic condition which can be detected and/or stored. In the computer memory, the "one" or "zero" is the presence or absence of charge, and millions of tiny capacitors in its memory store millions of "ones" or "zeroes." On storage media (storage discs, tapes, etc.), the "ones" or "zeroes" are represented by the alignment of magnetic particles, and on compact discs as the presence of optically-detectable characteristics.

A 2-based number system operates in a fashion which is analogous to our more familiar 10-based (decimal) system. The numbers we use every day are really representations of various powers of 10. For example, the number 24,365 in our normal 10-based system is really a summation of five numbers, with each digit representing a number times 10 raised to a certain power, as shown in Figure 12.10. Each digit can range from 0 to 9 (the base 10 minus 1).

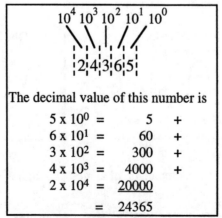

$$10^4 \ 10^3 \ 10^2 \ 10^1 \ 10^0$$

$$2 \ 4 \ 3 \ 6 \ 5$$

The decimal value of this number is

$5 \times 10^0 =$	5	+
$6 \times 10^1 =$	60	+
$3 \times 10^2 =$	300	+
$4 \times 10^3 =$	4000	+
$2 \times 10^4 =$	20000	
$=$	24365	

Figure 12.10 The meaning of the number 24,365 in a 10-based (decimal) system

The 2-based (binary) number system used by computers operates the same way as a 10-based system, except that each digit can range from 0 to 1 (the base 2 minus 1), and each digit position represents 2 raised to a certain power, as shown in Figure 12.11.

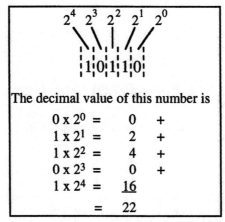

The decimal value of this number is

$$0 \times 2^0 = \quad 0 \quad +$$
$$1 \times 2^1 = \quad 2 \quad +$$
$$1 \times 2^2 = \quad 4 \quad +$$
$$0 \times 2^3 = \quad 0 \quad +$$
$$1 \times 2^4 = \quad \underline{16}$$
$$= \quad 22$$

Figure 12.11 The meaning of the number 10110 in a 2-based (binary) system

Some computers and related equipment can express a larger set of numbers than can other equipment. This is because all such equipment represents numbers in the binary system as a string of digits or "bits," and the length of the string depends on the design of the equipment. Therefore, hardware which is rated for 8 bits handles all numbers as strings of 8 digits. The largest number it can represent is 11111111, which in our decimal system equals 255 ($=2^8$-1). For 12-bit hardware, the largest number that can be represented is 4,095 ($=2^{12}$-1), and for 16-bit hardware it is 65,535 ($=2^{16}$-1).

The number of bits used to represent a value in an A-to-D converter or D-to-A converter (as described above) affects the "resolution" of that device, i.e. the ability of that device to express small differences between numbers. This is because a signal (e.g. 0-10V) sent through an A/D converter or D/A converter is represented by the full range of numbers of that converter. For example, in an 8-bit converter and a 0-10V signal, the 0-10V range is divided into 255 discrete values. Thus, the smallest difference that the device can represent is 10V/255 = 0.04V. On the other hand, a 12-bit device divides the 0-10V range into 4,095 values and could represent differences of 10V/4095 = 0.002V. Similarly, a 16-bit converter could resolve the signal into 65,535 values and could represent differences of 10V/65535 = 0.00015V. Obviously, for the most accurate measurement and adjustment, you would want an A/D converter or D/A converter with the highest number of bits.

We are finally ready to understand the hardware we need for the control loops of the acid-neutralization process. As a simple example, we decided to use feedback control to keep the liquid level in the HCl holding tank within tolerable limits. In that control loop, a level sensor (i.e. a level transducer) would monitor the level in the tank and convert it into a standard current or voltage signal, which would be sent back to a computer (Figure 12.12). The computer would need to have a card with an A/D converter to read the signal and convert it to digital form. The computer would need software which would compute a correction for the flow rate of HCl from the tank. A D/A converter would typically be needed to change that correction to a standard analog current or voltage signal. Further, that signal would need to be sent to the control valve in the HCl outlet line. Finally, the valve would need to include electronics and associated hardware which allowed its position to be governed by the current or voltage received from the computer and would then respond to the correction. With all of this in place, the control loop for the level would be operative.

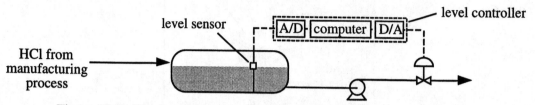

Figure 12.12 Hardware for controlling the level of HCl in the holding tank

With the rapid advances in computers and computer-related hardware, faster and faster computers and interface electronics are becoming available and are smaller and less expensive than ever before. One example is the presence of a computer and appropriate A/D and D/A converters on a single chip in modern automobiles (see Reading Question #3). In chemical processing, faster computers will also mean that process control will become ever more sophisticated and effective. This is a significant advancement in the chemical process industry, because it will lead to higher quality of products, less waste of materials, a safer processing plant, and better protection of the environment.

READING QUESTIONS:

1. Why is a combination of feedback and feedforward control better than using either feedback or feedforward control alone?

2. Based on what you have read about analog and digital signals, identify an analog clock or watch and a digital clock or watch. Explain the features which led you to make each identification.

3. Indicate whether an A/D converter or a D/A converter would be needed for each of the following:

   a. a computer (on a chip) in your car senses the concentration of carbon monoxide in the exhaust

   b. the same computer chip as in part "a" adjusts the air intake into the engine

HOMEWORK PROBLEMS:

1. For each of the following systems identify the control mode as either feedback or feedforward and identify the controlled variable and the manipulated variable.

   a. thermostat in your home or apartment

   b. speed (cruise) control in an automobile

2. Each of the following activities has some aspects which exhibit feedback control and some which exhibit feedforward control. For each activity, describe at least one aspect which is feedback and one which is feedforward.

   a. riding a bicycle

   b. being a student

3. Consider a feedforward control loop to control the cooling-water flow rate in the acid-neutralization process.

   a. What disturbances might be important to anticipate in the controller?

   b. Design (sketch) such a control loop and identify the important components.

   c. If you could only use a feedback loop or a feedforward loop for this particular application, which is preferred? Why?

4. A reactor in which an endothermic reaction takes place is held at constant temperature via a heating fluid circulated through a heating jacket around the reactor, as shown.

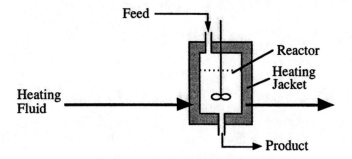

   a. Sketch a feedback control loop for this process to control the temperature inside the reactor by adjusting the flow rate of the heating fluid. In your sketch, label all the components of the loop.

   b. Sketch a feedforward control loop for this process to control the temperature inside the reactor by adjusting the flow rate of the heating fluid. Your control loop should be capable of responding to variations in the temperature of the incoming heating fluid and variations in the flow rate of the incoming reactor feed (and, hence, the rate of reaction).

5. Given an electronic system with a resistance of 250 *ohms*, how much current would flow through the system if 2.4 *volts* were placed across the system? Hint: 1 *volt* = 1 *ampere•ohm*

6. What is the equivalent decimal number of each of the following binary numbers:

       00111001
       110001110101
       1000000000000001

7. With a 32-bit A/D converter,

   a. How many values can be represented?

   b. What is the smallest voltage difference that can be represented from a 0-10V signal?

Courtesy of Bran+Luebbe, Buffalo Grove, IL

CHAPTER 13

ECONOMICS
(IS IT ALL WORTH IT?)

Up to this point we have focused principally on the technical aspects of engineering. However, most practical engineering decisions must be evaluated based on economics. Therefore, the ability to perform an economic analysis is an essential engineering skill. In this chapter, we identify different types of engineering costs and develop some of the basic skills needed for economic evaluation.

COSTS

There are two basic types of engineering costs, *capital costs* and *operating costs*. Capital costs are the initial costs of building the process, i.e. the costs associated with purchasing/building such items as reactors, heat exchangers, computers, control valves, piping, etc. Operating costs are the day-to-day costs of operating equipment and running a chemical plant. They include the costs of raw materials, pumping, heating and cooling, labor, maintenance, etc. It is important to consider both types of costs over the entire lifetime of equipment to get an accurate picture of overall cost. To illustrate this concept, let's consider the choice between two water heaters, one an expensive high-efficiency model and the other a less-expensive model with a lower efficiency, as shown in Table 13.1. In this example, the savings in the energy cost (operating cost) for the high-efficiency model over the ten-year lifetime of the water heaters more than compensates for the initial difference in the capital cost.

Table 13.1. Capital and Operating Costs for Water Heaters

	High-Efficiency Water Heater	Standard Water Heater
Purchase (Capital) Cost	$550	$338
Operating Cost per Year	$129	$172
Lifetime	10 years	10 years
Total Cost	$1,840	$2,058

There are several different types of costs associated with the design and construction of chemical processing facilities which are typically grouped under the general heading of *capital costs* or *capital investment*. These include the purchase, delivery, and installation of equipment items which were designed using the methods introduced in previous chapters (e.g. reactors, heat exchangers, etc.). The cost of piping, instrumentation, buildings, service facilities and land are also considered part of the capital investment. Also included are the costs of engineering and design, construction, contractor fees, and permits. You will learn more about these costs and how to estimate them in subsequent chemical engineering courses.

In this introductory treatment, we will use a simplified approach, useful for initial design calculations, to estimate the capital investment. We begin by determining the purchase cost of the individual equipment items which make up the chemical process. The most accurate way to obtain a purchase price for a particular equipment item is simply to call the vendor. However, general cost correlations are frequently used in place of vendor quotes since they are quicker and easier to use for initial design calculations. Examples of such correlations[1] are provided in Table 13.2. These particular correlations assume carbon steel as the material of construction and a base year of 1987. The Marshall & Swift Equipment Cost Index (M&S Index) reflects the current average cost for equipment and is inserted into the cost correlations to determine costs applicable to a particular year. For example, insertion of the M&S index for 1996 will provide a cost applicable to that year. Values for the M&S Index are published in several chemical engineering-related magazines (e.g. in the back of each issue of the magazine *Chemical Engineering*[2]).

Table 13.2 Sample Correlations to Estimate Equipment Prices (from reference 1)

$$\text{Tank or Reactor Cost (\$)} = \left(\frac{M\&S}{814}\right)\left(47.0\ V^{0.61}\right)$$
where V is the volume in *gallons*

$$\text{Heat Exchanger Cost (\$)} = \left(\frac{M\&S}{814}\right)\left(398\ A^{0.65}\right)$$
where A is the area of the heat exchanger in ft^2

$$\text{Centrifugal Pump Cost (\$)} = \left(\frac{M\&S}{814}\right)\left(421\ \dot{V}^{0.46}\right)$$
where $\dot{V}$ is the volumetric flow rate in *gal/min*

The purchase prices from Table 13.2 do not include delivery costs which are approximately 4 to 10% of the purchase price. As a conservative estimate, we will assume a delivery cost of 10% for our purposes here.

Once the cost of the equipment is known, it can be used to estimate the capital investment by multiplying the cost of the purchased equipment (delivered) by a scaling factor called a "Lang" factor. The value of the Lang factor for a major plant addition to an existing site has been estimated[3] to be *5.69*. Using this factor and the price of the delivered equipment, the capital investment can be estimated as shown in the following example:

Example 13.1

The purchase prices of equipment needed for a plant expansion total $900,000. Use the guidelines for delivery costs and capital investment given above to estimate the capital investment for this expansion.

Solution:

Purchase Price (given): $900,000

Delivered Purchase Cost: (110%)($900,000) = $990,000

Capital Investment: (5.69)($990,000) = $5,630,000

Operating costs include the costs of raw materials, utilities, electric power, labor and benefits, legal services, maintenance, advertising, public relations, insurance, transportation, etc. Some of these items can be directly calculated from prices of materials and utilities. Others are estimated from simple formulas based on experience. Detailed determination of the operating costs is beyond the scope of this book. For problems which we will consider, specific information will be given which will allow estimation of the operating costs for each particular problem.

PROFITABILITY

A task frequently faced by a design engineer is the evaluation of the profitability of a potential investment. The first task is to estimate the *gross annual profit*, which includes the *sales* per year, i.e. the number of items sold times the price per item. Subtracted from the annual sales is the yearly *operating costs*. Also subtracted from the sales is the *depreciation* per year of the facility, which reflects the fact that the value of the equipment, etc., decreases with time much as your car decreases in value. Thus the annual depreciation is approximated as the capital investment (or some fraction thereof) divided by the lifetime of the facility (in the USA, that "lifetime" is established by the Internal Revenue Service). Hence,

$$Gross\ Annual\ Profit = Sales - Operating\ Costs - Depreciation \qquad (13.1)$$

The profit after taxes is obtained by multiplying the gross annual profit by $(1-\phi)$, where ϕ is the fractional tax rate (e.g., if the tax rate is 35%, ϕ is 0.35). Therefore, the *Net Annual Profit After Taxes (NAPAT)* is:

$$NAPAT = (1-\phi)(Gross\ Annual\ Profit) \qquad (13.2)$$

To estimate profitability, the *NAPAT* is compared with the size of the initial investment and with the amount of profit that could have been generated by other kinds of investments with the same amount of money. One simple measure of profitability used in the past is the *Return On Investment (ROI)* which is defined as

$$ROI = \frac{NAPAT}{Capital\ Investment} \qquad (13.3)$$

and represents the fraction of the initial investment returned each year. It should be pointed out that, although it is suitable for our present purposes, *ROI* is no longer typically used by practicing engineers. This is because it does not account for the "time value of money" or the fact that it is more valuable to have a certain amount of money now then it is to have the same amount of money available at a future date. The time value of money is an important factor in the assessment of plant profitability since the money is invested now in order to make a profit in the future. You will learn how to account for this factor in later chemical engineering courses. For now, we will use the *ROI* to perform a preliminary assessment of profitability.

Investment in chemical plants or processes involves a certain amount of risk. This risk is greater than that associated with investments such as savings accounts or government bonds. For example, an unforeseen shift in the market demand can make an investment unprofitable. Consequently, a higher rate of return is required by investors to justify the additional risk. In other words, investors will not be willing to take the additional risk unless the potential return is significantly higher than that offered by alternate investments. Therefore, our company might not consider an investment to be justifiable unless it has an *ROI* value of approximately 15% or greater.

ECONOMICS OF THE ACID-NEUTRALIZATION PROBLEM

As mentioned above, the type of calculation described here applies to the modification of existing plants, such as the addition of a facility to conform to environmental regulations. Such modifications are evaluated by the companies involved to determine whether the costs can be justified, i.e. whether the overall plant and operation constitute a profitable venture as compared with alternate investments. If the conclusion is affirmative, the modifications are made. If the analysis yields a negative result (i.e. the profitability of the modified plant and operation are judged to be inadequate), the modifications are not made. In the latter case, the plant might continue to operate without the modifications if possible and if profitable. Conversely, if prohibited from operating by governmental restrictions or by other limitations, the plant might be closed.

In the case of our problem, the company needs to determine the profitability of the overall operation with the addition of the proposed acid-neutralization facility. The capital investment will be the cost to build the new facility. The new operating costs will include the cost associated with operating the existing plant (of course, without the expense of having the external contractor dispose of our acid byproduct) plus the costs of operating the new acid-neutralization process. The sales will continue to be generated by the product from the existing part of the plant, whose continued operation depends on the addition of the new facility. We will begin by estimating the capital costs for the acid-neutralization facility.

The capital costs associated with our acid-neutralization process will include costs for the following equipment items:

1. HCl tank
2. NaOH tank
3. HCl pump
4. NaOH pump
5. Cooling water pump
6. Heat exchanger

The purchase price of each of these items will now be estimated, and for that estimation, we will use the annual 1995 value for the M&S Index (for illustration purposes), which is 1027.

We have not discussed the size of the tanks for holding the HCl and NaOH. However, we learned in Chapter 3 that the average HCl flow rate would be 11,600 *L/hr* (51.1 *gal/min*), and we determined in Chapter 6 that the optimum flow rate for the NaOH would be 6,500 *L/hr* (28.6 *gal/min*), where it is left to you to verify the values in alternate units shown in parentheses. The sizes of the holding tanks would depend on such factors as the NaOH delivery rate and reliability, the average deviations in HCl flow rate from the existing process, etc. For our economic analysis, we will assume that these factors lead us to decide that our holding tanks should be able to hold one week's worth of HCl and NaOH, respectively. Thus, the volumes of the tanks would be:

HCl tank: $\quad \dfrac{11,600\,L}{hr}\left(\dfrac{0.26417\,gal}{L}\right)\left(\dfrac{24\,hr}{day}\right)\left(\dfrac{7\,days}{week}\right) = 514,814\,gal$

NaOH tank: $\quad \dfrac{6,500\,L}{hr}\left(\dfrac{0.26417\,gal}{L}\right)\left(\dfrac{24\,hr}{day}\right)\left(\dfrac{7\,days}{week}\right) = 288,474\,gal$

and, from Table 13.2, the estimated purchase prices (rounded to the nearest $100) of these tanks would be

$$\text{HCl tank:} \quad Cost \ (\$) = \left(\frac{1027}{814}\right)\left[47.0 \ (514,814 \ gal)^{0.61}\right] = \$180,800$$

$$\text{NaOH tank:} \quad Cost \ (\$) = \left(\frac{1027}{814}\right)\left[47.0 \ (288,474 \ gal)^{0.61}\right] = \$127,000$$

Also from the table, the purchase prices of the <u>pumps</u> are estimated directly from the flow rates. We note that the flow rate of required cooling water was determined in Chapter 10 to be 43,440 L/hr (191.3 gal/min). Thus, the purchase prices are:

$$\text{HCl pump:} \quad Cost \ (\$) = \left(\frac{1027}{814}\right)\left[421 \ (51.1 \ gal/min)^{0.46}\right] = \$3,200$$

$$\text{NaOH pump:} \quad Cost \ (\$) = \left(\frac{1027}{814}\right)\left[421 \ (28.6 \ gal/min)^{0.46}\right] = \$2,500$$

$$\text{Cooling water pump:} \quad Cost \ (\$) = \left(\frac{1027}{814}\right)\left[421 \ (191.3 \ gal/min)^{0.46}\right] = \$6,000$$

Finally, the required area for the <u>heat exchanger</u> was determined in Chapter 10 to be 89.3 m^2 or 818 ft^2, which allows us to estimate the purchase price from Table 13.2:

$$\text{Heat Exchanger:} \quad Cost \ (\$) = \left(\frac{1027}{814}\right)\left[398 \ (818 \ ft^2)^{0.65}\right] = \$39,300$$

The total purchase price of the equipment comes to $358,800. Thus, the complete estimate of the capital investment for the acid-neutralization process would be:

Purchase Price (as just calculated): $358,800

Delivered Purchased Cost: (110%)($358,800) = $394,700

Capital Investment: (5.69)($394,700) = $2,246,000

We will assume that the original plant has been fully depreciated, so we can evaluate the entire plant from this point in time based on this new investment. Suppose that we can depreciate the addition over 10 years (i.e. $225,000/year). Suppose, further, that a detailed analysis of the operating costs for the entire plant, including the new facility, yields an estimate of $3,827,000 per year. Gross sales are estimated to be 4.83 million dollars per year, and the tax rate for the company (including tax credits for environmental improvements) is 33%. Thus:

(Eq. 13.2): $NAPAT = (1-.33)(\$4,830,000 - \$3,827,000 - \$225,000) = \$521,000$

(Eq. 13.3): $ROI = \dfrac{\$521,000}{\$2,246,000} = 23.2\%$

Provided that this rate of return is considered justifiable by your company, the modification can receive approval. The only step now remaining is to formulate your proposal to your supervisors and to make a convincing presentation.

REPORTING THE RESULTS

In the design of a process or other project, the economic analysis is usually the final step, since it requires that all other aspects of the project be completed first. Therefore, completion of that economic analysis usually means that the project is at a point where the results can be communicated to other interested parties. This communication is usually in the form of a written report.

The most common formats for written engineering reports are *formal reports* and *short* or *memo reports*, and the choice of formats is based on the intended audience and on company preferences. Regardless of the format, engineering reports should communicate the following kinds of information:

1. summary of the problem (never assume that the boss or other readers remembers or knows what it was all about)
2. summary of the options that were considered and the option that was chosen
3. description of the process that was chosen and designed (including the major items of equipment)
4. summary of the economics of the process (including the capital investment and the ROI)
5. recommendations concerning whether to implement the design (i.e. to complete a detailed design and to build the facilities)

The longer formal report usually includes several sections, often with each one dedicated to one of the items on the above list. It usually also includes an abstract (to enable the busy executive to quickly read the most essential points of the report) and an appendix containing minute details of the procedures and analyses (to enable a colleague to reproduce or evaluate the work). The memo report incorporates all of the information in the above list but does so in a brief (often a one-page) presentation, usually with an extensive appendix attached. An example memo report for our acid-neutralization project is presented below.

ABC Chemical Company
Memorandum

To: Barbara Magelby, Supervisor, Chemical Process Group
From: <your name>, Project Engineer

This memo is to provide an update on the progress our group has made toward finding a solution to the HCl disposal problem. As you know, it is anticipated that the company that has been disposing of our HCl byproduct will soon be out of business. Our group has been charged to propose a strategy, design, and preliminary cost analysis for safely and legally disposing of the acid waste. To this end, we have considered several options including: 1) changing our company process so that the waste acid stream is not produced, 2) contracting with another independent company to dispose of the acid, 3) long term storage of the acid on site, 4) use of an evaporation pond to concentrate the waste solution, 5) treatment of the waste stream (acid neutralization) followed by discharge into the lake adjacent to the plant site and 6) closing the plant. Our initial analysis indicated that the most economical and reliable of these options would be to neutralize the acid ourselves and then dispose of the stream (Option #5). This conclusion is supported in the Appendix which accompanies this memo.

In support of this option, we have completed the preliminary design of an acid neutralization process for treatment of the waste stream. The process consists of storage tanks for the waste acid and the base (NaOH) that will be used to neutralize the acid, a mixer to facilitate mixing of the acid with the base, a heat exchanger to cool the stream to an environmentally acceptable outlet temperature, and the necessary pumps, piping, and instrumentation. The estimated capital cost for the project is $2.25 million. This option would allow the plant to continue operation at the current rate which produces $4.83 million in gross sales annually. The annual operating cost for continued production, including the acid neutralization process, is estimated as $3.83 million, where the current cost of acid disposal has been credited. The estimated ROI is 23.2%, well above the company minimum of 15%.

Given the above, we strongly recommend that the company pursue construction of the acid neutralization process as quickly as possible. We estimate that the project can be completed in less than a year and be on-line when needed. In case of delays, or if the facility is needed sooner than anticipated, it will be necessary to temporarily store the acid on-site. The treatment facility has been sized to accommodate acid flows in excess of those produced by our current process, so that the acid which must be stored can easily be treated at a later date. The excess capacity also provides the company with the option of increasing production rates in the future if desired. Additional costs associated with delays, etc. have been factored into the cost analysis.

Please do not hesitate to contact me if you have any questions or require additional information. We are ready to complete quickly the final design and begin construction should the company chose the recommended option. In any case, we await your instructions.

References

1. Baasel, W.D., *Preliminary Chemical Engineering Plant Design*, New York: Van Nostrand Reinhold, 1990.

2. *Chemical Engineering*, publication of McGraw-Hill Co.

3. Peters, M.S. and K.D. Timmerhaus, *Plant Design and Economics for Chemical Engineers*, Fourth Edition, New York: McGraw-Hill, 1991.

READING QUESTIONS:

1. What would the total cost for the two water heaters in Table 13.1 have been if figured for a lifetime of 5 years instead of 10 years?

2. a. Define each of the following: *NAPAT, ROI,* ϕ

   b. Describe in your own words what the *ROI* represents. Is it possible for the *ROI* to be greater than 100%? Justify your answer.

HOMEWORK PROBLEMS:

1. Your assignment is to determine the return on investment for an ammonia plant which is being considered for construction. The plant will produce 55,650 *kg/hr* of ammonia which sells for $0.60/*kg*. The plant will operate 24 hours per day and have a service factor of 0.92 (in other words, the plant will be producing during 92% of the available hours of the year). The total investment for the plant is projected as $166 million, which can be depreciated over 12 years. The annual operating cost of running the plant is estimated to be $220.4 million.

   a. Use a spreadsheet to determine the ROI for tax rates ranging from 0% to 50% (in 5% increments).

   b. What is the ROI for the current tax rate of about 35%?

   c. How significant are taxes when determining profitability? (please justify your response)

2. Determine a 1994 delivered purchase price for the items listed below. The M&S index for 1994 is 993.

   a. A tank with a volume of 500,000 *gallons*.

   b. A heat exchanger with an area of 3500 *ft²*.

   c. A centrifugal pump for pumping 150 *gal/min*.

CHAPTER 14

CASE STUDY
(INTEGRATING IT ALL TOGETHER)

This chapter consists of a design problem in which you will be asked to design a process for the production of xylenes. We hope that you will accomplish several things as you complete this design. First of all, the design problem will provide an opportunity for you to review many of the topics covered during the course of this semester. Second, the design will provide an opportunity for you to work in teams on a project. Finally, the design will provide the opportunity for you to apply the topics from separate chapters in an integrated fashion to the solution of a single engineering problem.

The Problem

Xylenes are used as raw materials for the manufacture of polyesters which are used for textile fibers, photographic film, and soft-drink bottles. The process to be added to an existing facility involves an isomerization reactor where meta-xylene (m-x) is converted to ortho-xylene (o-x) and para-xylene (p-x). Note that all three of these compounds are isomers (same elemental composition with different chemical structure). Therefore, they have the same molecular weight ($MW = 106$). For our purposes here we will assume that equal amounts of o-x and p-x are produced so that the reaction proceeds as follows:

$$2\text{m-x} \rightarrow \text{o-x} + \text{p-x}$$

This liquid phase reaction is assumed to be irreversible and is approximated by first-order kinetics with a rate constant of $k_r = 0.133\ min^{-1}$. The reaction is slightly endothermic ($\Delta \tilde{H}_{react.,m-x} = 295\ cal/gmol\ of\ m\text{-}x\ reacted$). The simplified process is described in the following paragraph.

A liquid feed stream will enter the process at atmospheric pressure and a temperature of $77\,^{\circ}F$ at a rate of one million kg/day. The mass fractions of that feed stream are:

Feed Stream	o-x	m-x	p-x
	0.29	0.48	0.23

A centrifugal pump will be used to increase the pressure of the stream sufficiently to overcome the pressure drop due to friction in each of the components of the process (see the additional information given below). Following the pump, the stream will pass through a shell-and-tube heat exchanger where the temperature will be increased to $500\,^{\circ}F$ using steam. The hot pressurized stream will then be fed into a well-mixed isothermal reactor (maintained with a steam jacket around the reactor) where 70% of the m-x will be reacted to form products. The product stream from the reactor will be cooled as much as possible with cooling water in another exchanger before entering the separation system. The separation system will yield three streams with the following mass fractions and which are all at the same temperature:

Output Stream	o-x	m-x	p-x
1	1.00	0.00	0.00
2	0.00	0.03	0.97
3	0.00	0.70	0.30

Technical Information

1. The heat capacity $(\overline{C}_p)$ for all xylene streams in this problem is essentially the same and can be approximated for this preliminary estimate by an average value of *35 Btu/lbmol°F*.

2. Assume countercurrent, single-pass heat exchangers for this project.

3. Cooling water is available at *90°F* and has a maximum return temperature of *120°F*. In selecting the actual return temperature, assume that you want to minimize the cost of cooling water (even though this will affect the size and cost of the heat exchanger). Also, the heat capacity of water is 1.0 *cal/g°C*.

4. In selecting the temperature of the process stream leaving the cooler, observe the rule of thumb that the minimum temperature difference (either ΔT_1 or ΔT_2, see Figure 10.11) for a heat exchanger is *10°F*.

5. Saturated steam is available at *545°F* (1000 *psig*), and $\Delta \hat{H}_{vap}$ = *650 Btu/lb_m*. When used for the heater and the reactor, the amount of steam which is used is only that which is condensed, and it leaves as saturated liquid.

6. Assume that the separation system is isothermal (operates at constant temperature).

7. The separation is difficult because of the close boiling points of the reactant and products and must be accomplished using a complex scheme of several steps. For this preliminary design, simply represent the separation scheme by a box labeled "Separation System."

8. The density of xylenes is given in Figure 14.1.

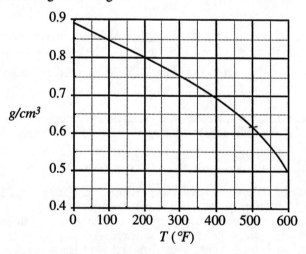

Figure 14.1 Density of xylenes

9. The pressure drop due to friction in each of the units is estimated to be as follows:

Unit	Pressure drop (*psi*)
Feed heater	15
Reactor	5
Product cooler	15
Separation system	165

The outlet pressure from the separation system is atmospheric pressure. For this preliminary design, the pressure drops in the pipes between the major components may be neglected. The pipes are all the same diameter, and there are no significant elevation changes. You may also ignore the changes in kinetic energy associated with having one inlet and three outlets in the separator (it would be instructive for you to think about how large such kinetic energy terms would be, say with a 6-inch ID pipe, compared with the large pressure drop term for the separator).

10. The following costs are applicable:

> Market Costs:
> | | |
> |---|---|
> | feed stream: | $0.88/gal |
> | ortho-xylene product stream: | $0.22/lb_m |
> | para-xylene product stream: | $0.22/lb_m |
> | byproduct (70% meta-xylene) stream: | $0.12/lb_m |
>
> Utility Costs:
> | | |
> |---|---|
> | steam: | $5.20/1000lb_m |
> | cooling water: | $0.03/1000gal |
> | electricity: | $0.05/kW-hr |

11. The *direct capital costs* for the reactor, heat exchangers, and pump can be estimated from the formulas in Table 13.2 and using the current value of the *M&S* Index (look in the back of a current issue of the magazine *Chemical Engineering*[1]).

12. The total operating cost per year can be estimated as 1.5 times the sum of the feed cost, steam costs, cooling water costs, and electricity costs (excluding the separation system for which these values are not known). In other words:

Operating Cost = 1.5 (Feed Cost + Steam Cost + Cooling Water Cost + Electricity Cost)

13. The current tax rate is 35%; we are allowed to depreciate our equipment over 10 years; and our company requires a minimum *ROI* of 0.15. In our case, 85% of the capital investment can be depreciated.

14. The cost of the separation system can be approximated as $6.5 million (delivered price).

Technical Assignment

1. Draw a PFD for the process.

2. Perform material balances on the process, one unit at a time, to determine the component flows for each stream. ✓

3. Perform energy balances to determine the unknown temperatures of the streams and/or the heat duties for the heat exchangers and reactor. ✓

4. Calculate the amount of steam and cooling water needed for the heat exchangers and reactor. ✓

5. Design the reactor, i.e. determine the reactor volume needed to get the specified conversion. ✓

6. Determine the work required for the pump assuming that the pump is 85% efficient (only 85% of the energy added to the pump is transferred to the fluid).

7. Size the heat exchanger for heating the reactor feed stream using an approximate value of U_o from Table 10.5.

8. Size the heat exchanger for cooling the reactor product stream using an approximate value of U_o from Table 10.5.

9. Complete the following stream table   -22.333 c)   (- 17476.8564

Stream Table

	Process Feed	Pump Outlet	Reactor Feed	Reactor Outlet	Separator Feed	Separator Outlet #1	#2	#3
Flows: (lb_m/hr)							9662	12233.8
m-xylene	14,000	—	—	4198 3.287	—	4983.28		
o-xylene	26,583⅓	—	—	13300.9945	—	41983287	—	—
p-xylene	21,083⅛	—	—	36,483.362467	—	—	312403	5243
Total	91,666⅔	—	—	41666.6	—	41983287	32,20	7447.8564
Temp. (°F)	770°F	770°F	500°F	500°F	1000°F	1000°F		147 psi
Pressure (psig)	147 psi	214.7						
	0	200	185	180	165		0	

10. Complete the following equipment specification list:

Pump:

  Required horsepower: _____

Reactor feed heater:

  Heat duty (Btu/hr): _____

  Area, A_o (ft^2): _____

  Required steam (lb_m/hr): _____

Reactor product cooler:

  Heat duty (Btu/hr): _____

  Area, A_o (ft^2): _____

  Required cooling water (lb_m/hr): _____

Reactor:

  Volume ($liters$): _____

  Heat duty (Btu/hr): _____

  Required steam (lb_m/hr): _____

11. Determine the ROI for this project and offer your evaluation concerning whether or not the project is justified economically.

12. Suggest three ways to potentially improve the economics of this process. Explain why you believe your suggestions may make a significant improvement.

Using Engineering Teams for This Case Study

The value of team engineering and the roles that participants can play in engineering teams were introduced in Chapter 3. This case study provides an opportunity for students to gain experience with this important aspect of engineering. Table 14.1 suggests some roles, particularly suited to team *learning* coupled with team *engineering* (see Reference 2), that might be used for this case study. As you begin your teams activities, you may want to refer back to Table 3.1 , which briefly described some of the stages of team development. With that reminder, you will want to help your team move past the "Storming" to the "Performing" stage as quickly as possible.

Table 14.1 Suggested Team Assignments for the Case Study

<u>Leader/facilitator</u>
    Identifies and explains goals
    Assembles/prepares materials for group sessions (runs for additional
        material, if needed)
    Communicates with the instructor (to reinforce this role, the instructor
        may want to require that only the leader communicate with him/her)
    Keeps the group focused and on track
<u>Technical expert/coach</u>
    Makes sure the work is technically correct
    Relates the current principles to material previously studied
    Ensures the correctness of summaries and student explanations
<u>Learning checker/promoter</u>
    Regularly checks each group member's understanding of the principles
        (this person might periodically ask each to explain procedures)
    Makes sure that all participate (no silent observers)
<u>Summarizer/recorder</u>
    Writes down the group's decisions
    Summarizes the accomplishments of the group session
    Edits the group's report
<u>Team observer</u>
    Keeps track of how well the group is cooperating

References

1. *Chemical Engineering*, publication of McGraw-Hill Co.

2. Johnson D.W., R.T. Johnson, and K.A. Smith, *Active Learning: Cooperation in the College Classroom*, Edina, MN: Interaction Book Company, 1991.

Courtesy of Conoco Inc., Ponca City, OK

Index